Mentoring in Acade

For other titles published in this series, go to
www.springer.com/series/7855

Mentoring in Academia and Industry
Series Editor: J. Ellis Bell, University of Richmond, Virginia

Biology is evolving rapidly, with more and more discoveries arising from interaction with other disciplines such as chemistry, mathematics, and computer science. Undergraduate and Graduate biology education is having a hard time keeping up. To address this challenge, this bold and innovative series will assist science education programs at research universities, four-year colleges and community colleges across the country and by enriching science teaching and mentoring of both students and faculty in academia and for industry representatives. The series aims to promote the progress of scientific research and education by providing guidelines for improving academic and career building skills for a broad audience of students, teachers, mentors, researchers, industry, and more.

Volume 1 Education Outreach and Public Engagement
by Erin L. Dolan

Volume 2 Active Assessment: Assessing Scientific Inquiry
by David I. Hanauer, Graham F. Hatfull, Deborah Jacobs-Sera

Volume 3 Getting the Most out of Your Mentoring Relationships
by Donna J. Dean

Donna J. Dean

Getting the Most out of Your Mentoring Relationships

A Handbook for Women in STEM

Donna J. Dean
Association for Women in Science
1200 New York Ave. NW
Washington, DC 20005
USA
donna_dean@comcast.net

ISBN 978-0-387-92408-3   e-ISBN 978-0-387-92409-0
DOI 10.1007/978-0-387-92409-0
Springer Dordrecht Heidelberg London New York

Library of Congress Control Number: 2009920679

© Springer Science+Business Media, LLC 2009
All rights reserved. This work may not be translated or copied in whole or in part without the written permission of the publisher (Springer Science+Business Media, LLC, 233 Spring Street, New York, NY 10013, USA), except for brief excerpts in connection with reviews or scholarly analysis. Use in connection with any form of information storage and retrieval, electronic adaptation, computer software, or by similar or dissimilar methodology now known or hereafter developed is forbidden. The use in this publication of trade names, trademarks, service marks, and similar terms, even if they are not identified as such, is not to be taken as an expression of opinion as to whether or not they are subject to proprietary rights.

Printed on acid-free paper

Springer is part of Springer Science+Business Media (www.springer.com)

Foreword

Mentorship practice has been part of the human experience since the Golden Age of Greece. Engaging with a mentor as a way to learn and achieve one's full potential is an ancient and respected practice. And, it has been the keystone on which the Association for Women in Science (AWIS) has built its program over the past three decades.

Trailblazers, such as Dr. Estelle Ramey and Dr. Anne Briscoe, experienced first-hand the isolation of women in the country's male-dominated scientific establishment and worked to build an organization that would promote women through mentoring relationships. Dr. Ramey, who earned her degree in physiology and biophysics and taught at Georgetown Medical School, was a well-known feminist speaker and writer. Noted for her great wit, she once quipped, "I was startled to learn that ovarian hormones are toxic to brain cells." Throughout her career, Dr. Ramey decried sexist comments and situations that treated women as less than fully human. She felt very strongly about how little, if anything, it took to extend a helping hand to someone else in a way that could really make a huge difference in her life. As she wrote in her book called *Letters to our Grandchildren*, "If I could leave you with any advice, it would be to speak words of caring not only to those closest to you, but to all the hungry ears you encounter on your journey through a cold world. Stop on the mountain climb to bring all those less lucky, less agile, or well endowed. It will make the view even more beautiful when you get to the top."

This dictum has served generations of women well and has been the leitmotif for AWIS programming since 1971. AWIS has a strong track record in providing mentoring as well as in authoring and publishing materials about mentoring such as our classic mentoring primer *A Hand Up: Women Mentoring Women in Science*. This new guide is designed specifically for protégés and builds on the collective wisdom within our organization.

Today, AWIS continues to serve women in all science and engineering disciplines by providing opportunities for networking, peer-mentoring, e-mentoring, coaching, and career enhancement around the country through local chapters and strategic partnerships. As the only national, multi-discipline organization for women in science, technology, engineering, and mathematics (STEM), our members include researchers, entrepreneurs, teachers, policy makers, and writers employed in industry, academia, government, and the non-profit sector. They span the range of seniority from undergraduates to postdocs to senior members of the National Academies of Science and of Engineering.

We invite you tap into this rich resource for self-discovery and personal development and wish you much success and happiness along your path.

Washington, DC   Janet Bandows Koster

Preface

Donna J. Dean, past-president and long-time member of the national Executive Board of the Association for Women in Science describes in the introduction to *Getting the Most Out of Your Mentoring Relationships* how her amazement at incredible scientific feats combined with her desire to use what she learned from her own experiences as a Ph.D. scientist to help guide others became the impetus for this handbook. In focusing on the needs of the protégé, she responds to one of the most pressing issues articulated by young women scientists. Women scientists seek mentoring not only in the early stages of deciding whether to major in science and what scientific field to choose, but also in later decisions about how to obtain promotion and advancement in their particular work sector and setting. In a study I undertook of 450 academic women scientists at research universities and small liberal arts colleges, "lack of camaraderie and mentoring due to small numbers of women" emerged as the issue cited second most frequently by women in all scientific disciplines as they plan their careers. The only issue mentioned more often and underlined as more significant was "balancing work with family responsibilities" (Rosser, 2004).

One woman scientist in that study articulated why mentoring by other women scientists is particularly critical: "Although possibly less now than before, women scientists still comprise a small proportion of professors in tenure-track positions. Thus, there are few 'models' to emulate and few to get advice/mentoring from. Although men could also mentor, there are unique experiences for women that perhaps can only be felt and shared by other women faculty, particularly in other Ph.D. granting institutions. Some examples of this: a different (i.e., more challenging) treatment by undergraduate and graduate students of women faculty than they would of male faculty; difficulties in dealing with agencies outside of the university who are used to dealing with male professors; difficulties related to managing demands of scholarship and grantsmanship with maternity demands. More women in a department would possibly allow a better environment for new women faculty members to thrive in such a department through advice/mentoring and more awareness of issues facing women faculty members." (respondent 26 in Rosser, 2004, p. 40)

Fortunately, as Janet Bandows Koster, Executive Director of the Association for Women in Science, underlines in her foreword, the organization has given mentoring a central priority in its mission and activities during the last three decades. Members of the Association for Women in Science (AWIS) concur that the most significant contribution they make to other women scientists is mentoring. In 2005, I conducted a survey of AWIS Fellows. Launched in 1996 as part of the 25th anniversary celebration for AWIS, the Fellows Program aims to recognize and honor women and men who have demonstrated exemplary commitment to the achievement of equity for women in science, technology, engineering, and mathematics (STEM). Most AWIS Fellows have achieved considerable success in their own career that has brought them to a position where they have enabled women in STEM at a level worthy of national recognition. The AWIS Fellows include a significant number of university presidents, CEOs of major corporations, executive directors of professional societies, foundations, or non-profit organizations, as well as deans, department chairs, professors, government agency heads, and industrial research scientists.

In response to the question, "In your opinion, what changes in institutional policies and practices are most useful for facilitating careers of academic women scientists or engineers at the junior level?" AWIS Fellows demonstrated eloquence in identifying problems for junior and senior women in both their overall careers and in the laboratory environment. The Fellows also had thoughts about the changes in institutional policies and practices that would be most useful for facilitating the careers of academic women scientists and engineers, particularly at the junior level. "Mentoring for junior faculty" emerged as the response most frequently given by AWIS Fellows (41.3 percent) as the institutional policy/practice most useful for facilitating the careers of junior academic women scientists or engineers:

> Intense, active, continuing mentoring and establishment of support groups (breakfast, lunch or dinner on a regular basis–i.e. weekly or biweekly, gatherings) where women feel comfortable airing their concerns, gripes, fears, questions, to get reassurance, information, advice...from their peers (Rosser 2006, p. 286).

Getting the Most out of Your Mentoring Relationships: A Handbook for Women in Science, Technology, Engineering, and Mathematics responds to the needs articulated both by junior women scientists for mentoring and identified by senior AWIS Fellows as a top priority for facilitating careers in science. The *Handbook* provides the type of advice mentors need to guide their protégés successfully, while also allowing young women scientists to understand their role as protégés. Thank you, Donna J. Dean, for writing this handbook to fill a needed gap in the mentoring literature, and thank you, the Association for Women in Science, for publishing this volume and continuing to support mentoring activities.

Atlanta, Georgia   Sue V. Rosser, Ph.D.

Acknowledgment

For those who are part of the network, the resource, and the voice for women everywhere in the fields of science, technology, engineering, and mathematics

Contents

1 How to Use This Guide 1

2 What Is Mentoring? 3

3 Preparing to be Mentored 7
   3.1 Identifying Your Mentoring Needs 7
      3.1.1 Acquiring the Requisite Professional Credentials 7
      3.1.2 Recognizing when a Rich Opportunity Arises 7
      3.1.3 Learning from Mistakes or Missteps 8
      3.1.4 Dealing with Own Biases and Misconceptions 8
      3.1.5 Developing a Sense of One's Career Directions and Timing 8
      3.1.6 Selecting Appropriate Role Models 9
      3.1.7 Meshing One's Values with the Workplace 9
      3.1.8 Balancing the Pieces of One's Life 9
      3.1.9 Creating Opportunities for Others 10
      3.1.10 Knowing When to Move On 10
      3.1.11 Calculated Risk Taking 10
      3.1.12 Points to Ponder 11
   3.2 Mentoring Models – The Right One at the Right Time 11
      3.2.1 Points to Ponder 15
   3.3 Techniques and Tools for Starting a Mentoring Relationship 15
      3.3.1 How Do I Pick a Mentor and Start the Process? 17
      3.3.2 What Questions Should I Ask? 17
      3.3.3 Are There Any "Rules" I Should Follow? 18
      3.3.4 Points to Ponder 19

4 Mentoring Relationships 21
   4.1 What Makes a Relationship Work 21
      4.1.1 Be Yourself and Do Well by People in All Your Interactions 21

	4.1.2	Never Embarrass Your Mentor or Put Your Mentor in an Awkward Position	22
	4.1.3	Look for Patterns in Your Life and in Your Career	22
	4.1.4	Have a Sense of Humor	22
	4.1.5	Recognize that Your Actions, Whether Good or Bad, will Often have Consequences	23
	4.1.6	Seek the Hidden, Unwritten, and Inside Rules	23
	4.1.7	Points to Ponder	23
4.2	Mentoring Impact: What Protégés Say		23
	4.2.1	What do Protégés Want?	24
	4.2.2	What do Protégés not Want?	24
	4.2.3	Points to Ponder	27
4.3	From Protégé to Mentor: Voices From the Field		27
	4.3.1	Questions Posed to Mentors	27
	4.3.2	Many Mentors, Many Perspectives	28
	4.3.3	Never Too Busy to Help	29
	4.3.4	Finding a Positive Place	30
	4.3.5	E-Mentoring to Attain Workplace Success	31
	4.3.6	Tapping into the Pipeline	31
	4.3.7	Mentoring Is Colorblind	32
	4.3.8	Power Neutral Mentoring	33
	4.3.9	Age Doesn't Matter	34
	4.3.10	Points to Ponder	35

5 Changing Dynamics, Changing Needs ... 37

5.1	Mentoring for Under-represented Groups		37
	5.1.1	MentorNet – The E-Mentoring Network Focused on Diversity in Engineering and Science	39
	5.1.2	Points to Ponder	40
5.2	Mentoring in Cyberspace		40
	5.2.1	Mentoring in the New Era of Social Media (Web 2.0)	41
	5.2.2	Using Web 2.0 Technology to Empower Specific Groups	43
	5.2.3	Science Careers Forum	44
	5.2.4	Ph.D. Career Clinic	44
	5.2.5	Points to Ponder	44
5.3	Life-long Mentoring		44
	5.3.1	At the Student and Trainee Level	45
	5.3.2	At the Postdoctoral Level	46
	5.3.3	At all Post-training Career Levels	48
	5.3.4	Points to Ponder	49

6	**Career and Life Transitions**		51
	6.1 Work–Life Balance		51
		6.1.1 Finding Time for the Other Things in Life	51
		6.1.2 Managing Your Employer's Expectations and Your Own	52
		6.1.3 Strategies to Attain Balance	53
		6.1.4 Recognize That Balance Is Not Always Attainable	53
		6.1.5 Points to Ponder	54
	6.2 Coaching or Mentoring? What Do I Need Now?		54
		6.2.1 Embracing New Communication Paradigms and Developing Priorities	56
		6.2.2 Points to Ponder	57
	6.3 Transitioning into a Different Career Pathway		57
		6.3.1 Know Your Strengths, Your Thought Processes, and Your Values	58
		6.3.2 Let Go of What You "Should" Want	58
		6.3.3 Make Your Scientific Background Work for You	58
		6.3.4 Put a Price Tag on Procrastination	59
		6.3.5 Create Your Opportunities	59
		6.3.6 Points to Ponder	59
7	**Navigating Interpersonal Contexts**		61
	7.1 Inappropriate Relationships with Mentors or Supervisors		61
	7.2 Identifying Problematic Behaviors		62
	7.3 Defining Issues		63
	7.4 Intervention Strategies		66
8	**Starting Out with the Right Education**		71
	8.1 Compelling Voices		71
	8.2 Informal Education		72
	8.3 Precollege Education		73
	8.4 Undergraduate Education		74
	8.5 Graduate Education		76
	8.6 The Postdoctoral Years		78
9	**Moving Toward Career Success**		81
	9.1 Making the Connections		81
	9.2 Leaping Barriers and Achieving Goals		83
	9.3 Timing and Choices		85
	9.4 Facing the Gender and Diversity Issues		87
	9.5 Staying the Course		93

10	Voices of Experience		95
	10.1	Women Speakers: Make the Most of Your Moment	95
		10.1.1 Memorize Your Introduction and Conclusion	96
		10.1.2 Talk to Your Audience	96
		10.1.3 Watch the Clock	97
		10.1.4 Use Visual Aids Carefully	97
		10.1.5 Practice, Practice, Practice	98
	10.2	Things your Professor Should Have Told You	98
		10.2.1 Gaining Opportunity, Equality, and Power	99
		10.2.2 Learning the Academic Structure	100
		10.2.3 Starting off Right	101
		10.2.4 Avoid a Common Pitfall	102
		10.2.5 A Word about Extracurricular Activities	102
		10.2.6 Be a Good Mentor	103
		10.2.7 Make Your Work Visible, Known, and Valuable	103
		10.2.8 Once in Power	103
	10.3	Applying for Fellowships or Research Grants	104
		10.3.1 Graduate School and Postdoctoral Fellowships	105
		10.3.2 Research Grants	107
	10.4	Keys to Success in Graduate School and Beyond	109
		10.4.1 Choosing an Advisor	110
		10.4.2 Joining a Lab	110
		10.4.3 Meeting With Your Thesis Committee	112
		10.4.4 Some Final Tips	112
	10.5	Building Confidence and Connection	113
		10.5.1 Sexism, Internalized Sexism, and Stereotypes about Scientists	113
		10.5.2 Claiming our Intelligence, Confidence and Femaleness	115
		10.5.3 Building Our Connections to Others	117
		10.5.4 Towards Gender Equality	118
	10.6	Helping Those Who Follow	118
		10.6.1 How Are Girls Being Guided?	119
		10.6.2 Opening Doors for Girls	120
		10.6.3 Thinking Globally	121
		10.6.4 How Well Are We Guiding?	121
	10.7	Professional Responsibility	123
		10.7.1 Literacy and Expertise	124
		10.7.2 Remembering the Big Picture	124
		10.7.3 Helping Women Entering Science	125
		10.7.4 Working With Our Colleagues	126
		10.7.5 Serving Society	126

11	**Provocative Thoughts for a Better Future**		129
	11.1	The 'Problem' of Women in Science: Why is it So Difficult to Convince People that There is One?.	129
		11.1.1 Patriarchy	129
		11.1.2 Upending Traditions	130
		11.1.3 Implications for Women in Science	131
		11.1.4 Youth and Genius	132
		11.1.5 The Third Gender	132
		11.1.6 In Pursuit of "Excellence"	133
		11.1.7 The Purpose: To Change Scientific Dogma	135
		11.1.8 The Benefits for Humankind	135
	11.2	Tacit Discrimination and Overt Harassment: The Toll on Women, Minorities and the Nation	136
		11.2.1 Postsecondary Science for Women: What Welcomes and What Inhibits	137
		11.2.2 Results from the Literature Review	138
		11.2.3 Results from Three University of Michigan Research Projects	138
		11.2.4 The Stacked Deck Against Women in Science	139
		11.2.5 Addressing the Odds for a New Workforce	140
		11.2.6 A Mandate for Change	141
	11.3	The Red Shoe Dilemma	142
		11.3.1 Choices	143
		11.3.2 Having It All? Hardly!	144
		11.3.3 A Dangerous Myth	145
		11.3.4 One Woman's Path	145
12	**Resources**		149
	12.1	National Organizations for and of Women in Science, Technology, Engineering, and Mathematics	149
	12.2	Mentoring Resources	149
	12.3	Organizations with Special Focus on Equity in STEM	150
	12.4	Organizations with Focus on Equity for Women	151
	12.5	Field Specific Resources	151
		12.5.1 Aerospace	151
		12.5.2 Agronomy	152
		12.5.3 Anthropology	152
		12.5.4 Astronomy	152
		12.5.5 Biology	152
		12.5.6 Biomedical Sciences	153
		12.5.7 Chemistry	153
		12.5.8 Computer Sciences and Information Technology	154
		12.5.9 Education	154

12.5.10	Engineering	154
12.5.11	Geography	155
12.5.12	Geosciences	155
12.5.13	Mathematics, Statistics, and Economics	155
12.5.14	Medicine and Health	156
12.5.15	Meteorology	157
12.5.16	Physics	157
12.5.17	Psychology	157
12.5.18	Sociology	158
12.5.19	Toxicology	158
12.5.20	Veterinary Medicine	158

References 159

Bibiliography 161

Index 163

Introduction

> *There was a wall. Like all walls, it was ambiguous, two-faced. What was inside it and what was outside it depended upon which side of it you were on.*
> – Ursula K. Le Guin, *The Dispossessed*, 1974

From the moment that I could read, I have been transfixed by the incredible scientific feats and discoveries of the protagonists in science fiction, whether for good or for evil. I was overjoyed, in the very first science fiction book that I read, by the strong and leading roles that females evidenced in these other places and times. I was mesmerized by the incredible creativity with which scientific principles were exploited by authors who could tell good tales that combined a strongly human story within a context of science. That many of these writers were themselves in scientific fields was my first encouragement as a pre-teen with strong interests in science and mathematics.

It was only much later, well after I was in graduate school, that I realized I was missing something critical in my pathway forward. While I could not describe or verbalize it, I now recognize that I was looking for a mentoring framework. I needed help in understanding those ambiguous walls before me, and guidance in perceiving what was "inside" and what was "outside" so that I could choose what was best for me. I had to struggle over many years to develop my own mentoring relationships for life and career. Now that I look back over my subsequent career, I believe that my own directions in STEM have been influenced by "random collisions" with other individuals who were there when I needed them for mentoring. There certainly did not seem to be a series of "ordered events" that made my particular pathway a logical or relatively smooth course to follow. Out of my own experiences, both good and bad, came the impetus for this handbook. It is intended to provide tools, techniques, and resources so that you, the reader, can be better prepared for those "random collisions" and "ordered events" which can enrich both your life and career work. As a quick, yet structured, guide to being mentored, it addresses finding the right mentors, being a good protégé, and making the most out of today's diverse mentoring environments.

Many women now in STEM careers do speak of pivotal events, or persons, that imbued them with the enthusiasm and desire to pursue their interests in a STEM field. Much has been accomplished in creating the frameworks, concepts, and approaches that mentors should take, but the primary focus has been on the mentor, not the protégé. As we take an active role in selecting appropriate mentors for ourselves, we can acquire useful advice in our career directions, learn how to mesh our personal values with the demands of the workplace, and know when it is time to move on to our next career step. By sharing our challenges with mentors, we can clarify our feelings, values, beliefs, and attitudes that positively (or negatively) influence our career trajectories. Is this going to be easy or will every woman's path be smooth? Assuredly not, but having good mentors provides venues for support networks, the sharing of perspectives, and the acknowledgement that even the most successful and accomplished female scientists have surmounted obstacles and have made tough choices throughout their careers. For me, the "glass ceiling, sticky floor" analogy is apt – the glass ceiling represents the barriers and obstacles that others create which may impede our scientific endeavors, but the sticky floor represents those things that we do (or fail to do) when the choices are ours to make.

I extend particular thanks to Monica Horvath, Jane Chin, and Marilyn Suiter, who have written excellent articles on pathways to success in STEM careers for the quarterly magazine of the Association for Women in Science. The critical role of knowing how to be mentored throughout one's career as a female scientist has been a central theme for each of them. They, along with many other individuals in the Association for Women in Science, have constituted a powerful and caring network, resource, and voice particularly for the preparation of Chapters 5, 6, 10, and 11 of this handbook. Bernice R. Sandler graciously permitted the inclusion of her strategies for addressing inappropriate behavior in Chapter 7. The concepts in Chapters 8 and 9 are adapted from research conducted by the Association for Women in Science under a grant from the National Science Foundation (PGE Grant No. 020865), which comprised a part of the book for mentors entitled *"A Hand Up: Women Mentoring Women in Science"*.

A particular quote from Douglas Adams (author of *The Hitchhiker's Guide to the Galaxy* in 1979) has resonance in my own life and career: "I may not have gone where I intended to go, but I think I have ended up where I needed to be". I hope the same will be true for each of you.

Washington, DC   Donna J. Dean, Ph.D.

Chapter 1
How to Use This Guide

This book is structured to provide overviews and perspectives on a variety of topics and issues relating to being mentored in the diverse fields encompassed by science, technology, engineering, and mathematics (STEM). It is not meant to be read cover to cover, but rather to serve as a handbook in times of need when a few moments of reflection and perspective would be highly useful. Some will find many topics of interest and relevance, while others may find only a few items that guide them to a fuller grasp of what to expect in mentoring relationships.

There have been many superb books and articles written for mentors and from the mentor's perspective. Those seeking quality mentoring can clearly find these of significant assistance, but framing questions, concerns, and issues from the protégé's viewpoint creates better guideposts for action. Because this book is all about the protégé and not the mentor, four chapters are structured specifically to place mentoring in context. Put simply, these are as follows:

- **"What about me?"** Chapter 3 focuses on the "me-ness" of mentoring, in providing some frameworks for self-analytical thinking as protégés assess themselves and their needs.
- **"What about me and my mentor?"** Chapter 4 discusses the mentoring relationship from the perspective of how protégés can develop, enhance, and sustain productive interactions, with examples from real-life experiences.
- **"What about me and my life?"** Chapters 5 and 6 address the role of the protégé–mentor relationship in sustaining a meaningful life and career in the context of family, culture, and workplace change.

The remaining portions of the handbook (Chapters 7, 8, 9, 10, 11, and 12) provide a rich compilation of viewpoints, strategies, perspectives, and resources that address specific topics and concepts important to a woman's career in the STEM professions.

By delving into the various sections and vignettes of these chapters at any given point in time, readers can put together their own menu of helpful approaches and techniques. For life and career, these can be as follows:

- Ability to understand the forces that surround one and to develop essential conflict-handling skills
- Capacity to develop effective means of understanding and assessing organizational cultures, customs, and structures
- Knowledge of key external factors that society, traditions, and norms may control or strongly influence
- Acknowledgement of one's feelings, beliefs, values, and attitudes

Chapter 2
What Is Mentoring?

> *I can't say as ever I was lost, but I was bewildered once for three days.*
>
> Daniel Boone

In today's scientific enterprise, one must always be expanding one's knowledge set and skill base to remain competitive. Coupled with the extremely long training periods required to obtain many specialized scientific jobs, 'mentoring' has become almost a marketing buzzword to describe benefits implicit in professional organization memberships, graduate programs, and workplace environments. However, mentoring is a much more complex concept, with roots in ancient history and literature. In Greek mythology, Mentor, son of Alcumus, was, in his old age, a friend of Odysseus. When Odysseus departed for the Trojan War, he asked Mentor to be in charge of his son Telemachus and of his palace. When Athena visited Telemachus, she took the disguise of Mentor to hide herself from the suitors of Telemachus' mother Penelope. As Mentor, the goddess exhorted Telemachus to stand up against the suitors and to go abroad to learn out about his father.

According to Wikipedia, the first recorded modern usage of the term "mentor" can be traced to a book entitled *Les Aventures de Telemaque*, by the French writer François Fénelon, published in 1699. In that book, very popular during the 18th century, the lead character is Mentor. This frames the source of the modern use of the word mentor: a trusted friend, counselor, or teacher, usually a more experienced person.

Therefore, in summary, a mentor is a wise and trusted counselor or teacher. The object of the mentor's attention is hence called a mentee, meaning simply (and somewhat circularly) a person who has a mentor. However, a more appropriate term for the mentee is protégé. This term conveys not only the counsel of a mentor who is more prominent or influential, but also that the mentor is guiding, protecting, and promoting the protégé's career, training, and overall wellbeing.

Table 2.1

Mentor: a wise and trusted counselor or teacher
Protégé: one whose welfare, training, or career is promoted by an influential person (i.e. mentor)

Mentors are different from advisors, supervisors, and coaches – all terms encountered in the science, technology, engineering, and mathematics (STEM) fields and used interchangeably and often erroneously. An *advisor* is someone who offers unsolicited advice, though from a perspective of some amount of wisdom or authority. A *supervisor* is a person with the official task of overseeing the work of others. In addition, a term now used more frequently in scientific and engineering fields is coach, meaning a trainer or instructor, who helps with skill and ability development using didactic and experiential approaches. An important point is that a mentor is not by definition the Ph.D. advisor or postdoctoral supervisor, although many graduate and postdoctoral advisors/supervisors are mentors in the best sense of the term.

Table 2.2

If you acquire the right mentoring as a protégé, you will learn how to:

Control your attitude
Own your values
Practice good stress management techniques
Choose your battles carefully
Be tolerant of others' mistakes (and your own)
Keep your sense of humor
Avoid turning your strengths into weaknesses
Support peers and higher management, but not at the expense of your own principles
Behave consistently and decently
Build up trust with others
Understand your passions, values, and motivations
Trust your own experience but be open to feedback

Chapter 3
Preparing to be Mentored

3.1 Identifying Your Mentoring Needs

> The most important conversation you will ever have is the one you have with yourself.
> – Anonymous

Because none of us will have a perfect pathway to our career or to how that career will mesh seamlessly into our lives, we can learn and grow from the perspectives of others. A necessary first step to fruitful mentoring relationships is introspection that leads to better knowledge and understanding of oneself. While such introspection is an intrinsic part of one's journey through life and career, it can also be more painful and arduous at the times when most needed. To navigate through these complexities, it may be helpful to consider the following ten topics in assessing which aspects you want to work on. You may choose to prioritize them, to eliminate some of them, or to use them as a springboard to develop your own personalized list of mentoring needs at any particular point in time.

3.1.1 Acquiring the Requisite Professional Credentials

In most cases, acquiring a bachelor's degree in a STEM field is only the first step in preparation for a career track. Whether graduate school, a professional school such as law or medicine, or a few years in the workforce follow immediately upon graduation from college or university, there are usually basic credentials that must be acquired for further career growth. Identifying someone as a possible mentor from the sector or sectors of most interest to you will provide you with a sounding board and source of advice for whether that is indeed a pathway you wish to follow.

3.1.2 Recognizing when a Rich Opportunity Arises

You can become attuned to the patterns and career steps of people whose professional pathways you may wish to emulate. While serendipity often places opportunities (or obstacles) before us, these chance happenings often catch us off guard. Most mentors will be able to recount examples of missed opportunities missed and of opportunities taken. From this experiential base in the context of your own career pathway, you can begin to develop a sense of when you should step forward to seize the opportunity or step aside to deflect the opportunity (or obstacle). Particularly at the early stages of one's training and career, it is very possible not to recognize these 'signals' because we have not yet acquired sufficient experience, knowledge, or useful advice.

3.1.3 Learning from Mistakes or Missteps

Most practitioners in the STEM disciplines will acknowledge that, in the research or development laboratory, greater learning comes from the experiment or design that did not work the first time (or the second, or the third). While frustration or anger may be the immediate reaction, a more thoughtful and reasoned analysis can lead to a better experimental approach, an adjustment of incorrect design parameters, or a total rethinking of the concepts. Easy to do in the laboratory or design studio, but very difficult to apply the same principles when it is our own thinking or actions that are "not working". When mentors are open to sharing their own mistakes and missteps with us, we can learn from them not only WHAT but also HOW to acknowledge these and move forward. Be wary of the mentor (or potential mentor) who cannot acknowledge that she or he has made mistakes or missteps!

3.1.4 Dealing with Own Biases and Misconceptions

We all bring to our careers the context of our communities and families of origin. Sometimes, these can prevent us from seeing our own selves clearly. Many of us have been recipients of grossly incorrect perceptions such as "girls can't do math" or "women aren't cut out to be aggressive researchers". We should be diligent in assuring that we do not view others in similar ways. Often, a mentor from another culture, or discipline, or socioeconomic background can help us forthrightly confront and deal with these misperceptions.

3.1.5 Developing a Sense of One's Career Directions and Timing

Being attentive to the "clock" is an important facet of career and life growth, whether it is taking note of the average time to doctorate in your chosen discipline or to something as fundamental as when to become a parent. For most individuals, there is no one perfect time. In fact, the fluidity of careers is much greater now than previously. There are many more ways in which persons at all career stages are balancing personal life with career goals and life stages. And, accordingly more potential mentors are available who have grappled with the same issues. Having a mentor who can help you navigate the pathway to what is best for you could be one of the most critical mentoring functions.

3.1.6 Selecting Appropriate Role Models

The world of STEM would be a dull place indeed if everyone thought alike and lived their lives identically. As you embark on, or continue, your pathway to more complete self-knowledge, it is essential to acquire mentors who can assist you to in that part of the journey. Sometimes, a mentor may be very close to a role model of what you wish to be or to become. At other times, a mentor may be an "anti" role model for where and how you wish to pursue your career and life. This is by no means a bad thing. Often, we can achieve the best learning for ourselves from someone who is very different from us in philosophy, in lifestyle, and from career focus.

3.1.7 Meshing One's Values with the Workplace

It is critically important that one be in a workplace that is a reasonable match to one's values. For some people, this may be in a start-up biotechnology firm; for others, a government regulatory position; and for others, the academic arena. While the perfect job rarely appears before us magically and at the time we most want it, mentors from various sectors can lead us to a greater appreciation of the different organizational cultures. With that base of information, we are more likely to figure out (perhaps through an iterative job process) where we are most likely to be productive and generally satisfied.

3.1.8 Balancing the Pieces of One's Life

Knowing that there are only 24 hours in a day, and 7 days in a week, does not necessarily prepare us to deal with the choices we must make about how we allocate our efforts. Certainly as professionals in the STEM fields, we have made a commitment to our discipline and the job arena in which we are employed. But to be mentally and physically healthy human beings, we must also attend to ourselves and to our close others such as family and friends. Mentors will have made these "allocation" decisions, either deliberately or operationally by how they have structured their lives and career. Their perspectives can help us decide what is best for our own situation, at whatever stage of our lives. Time and effort devoted to those aspects that are not our work life may vary widely over decades, but we must decide our own best balance.

3.1.9 Creating Opportunities for Others

We can learn from mentors the ways in which more visibility, responsibility, and challenges can enhance career growth and open new doors of opportunity. They can provide examples and illustrations of where even the most seemingly minor event or activity has led to a new path forward for their protégés. An important aspect of being a "successful" protégé is to learn that, regardless of professional career level and stage, one can always foster opportunities for others. This does not mean establishing an expectation of direct benefit to oneself. However, sometimes indirect benefits may occur later as one's reputation as a collegial member of the research community becomes more widely recognized. It is important to remember that it is not necessarily the senior manager or the tenured professor who can provide opportunities. All of us are in a position to do this, regardless of what career stage or level we occupy.

3.1.10 Knowing When to Move On

One of the most valuable lessons in a career is to learn how to recognize that one's effectiveness and ability to grow in a job are at a standstill. For each of us, this will be different. For some, the signs may be obvious and recognizable with a bit of self-generated analytical thinking. For others, it may not be obvious at all. In these cases, mentors can provide needed perspective, "tough love", and other advice to help the protégé determine what is best overall. Sometimes moving on may mean seeking a different job in a different place; in other cases, it may merely mean developing a new attitude and perspective on ourselves and the workplace we occupy. And, there are many options in between. Failing to deal with this circumstance when it arises is never a good thing, but the ability to recognize the possibility of needing to "move on" is something that will be needed numerous times over the course of a career. Mentors can help us to assess what our options and reasonable choices are likely to be.

3.1.11 Calculated Risk Taking

Having chosen to pursue a career in the STEM fields already is an indication that one is comfortable with some measure of risk. While each of us occupies a position somewhere along the continuum between highly risk averse and highly risk seeking, it is important to be aware that stretching beyond one's comfort zone is a critical part of being a successful STEM professional. But, it is important to note that this risk-taking must be thoughtful and reasonably well considered, particularly if it involves the need to encounter or confront one's work colleagues. This is where the "calculation" enters; choosing to take risks is neither arbitrary nor spur of the moment, but rather reflects a careful analysis of the strengths and weaknesses of our strategy and the ultimate goal we wish to achieve. In this arena, the advice and counsel of appropriate mentors is invaluable. By having broader knowledge of the workplace or of the dynamics in the STEM field or of human nature, mentors can help protégés determine where, when, and how to undertake risks that are more likely to have some measure of success.

3.1.12 Points to Ponder

- Assess the key areas for which you seek mentoring.
- Consider your perspectives on work–life balance.
- Formulate a 'picture' of your ideal mentor.

3.2 Mentoring Models – The Right One at the Right Time

> As the years roll by, I find myself talking to people 30–40 years younger than myself. They have so much energy! And they know about some topics that I have not heard of. The age difference does not necessarily get in our way; in fact, sometimes it is helpful. We can learn from each other.
>
> Cathy Middlecamp

As one embarks on the pathway to a career in STEM, the young scientist needs to learn to balance the different roles that lie ahead in both career and life. These changing needs will influence the types of mentoring that are needed in the framework of long-term strategies for career planning. Mentors can be people who influence facets of your life and career positively during interactions that can last a lifetime or just a minute. Good mentors are savvy, consistent, honest, and patient. There may even be persons whom one considers as "anti-mentors"; that is, people with whom you have good relationships but who may not have the same priorities or values that are valid for you. However, one can learn important life and career strategies from such individuals by comparing and contrasting their perspectives with your own views and needs. In this circumstance, knowledge of one's self is an imperative so that choices can be made knowingly and in context.

Table 3.1

Ten Myths About Mentoring

1. Having a mentor is the best way to succeed.
2. Mentors should be older than protégés.
3. A close, intense relationship is the best primary way to learn about one's profession and to move up the ladder.
4. Mentoring relationships must be long lasting to be truly useful.
5. A person can have only one mentor at a time.
6. Mentoring is a one-way relationship, benefiting only the protégé.
7. Protégés must be invited to be mentored by the mentor.
8. When men mentor women, a sexual encounter is inevitable.
9. Men are better mentors for women.
10. The mentor always knows best.

Because professionals are more self-reliant, solutions to problems should be generated from within, but may benefit from the perspectives of mentors. In developing a strong and supportive relationship, mentors should not hesitate to ask the hard questions of protégés. And, in turn, protégés should recognize that dedicated mentors will help them uncover critical, though often painful, truths. As well, the roles of organizational culture and workplace diversity will exert an important influence on the mentoring needs of protégés. In some fields of endeavor it is likely that females might have different expectations about mentoring than males. All of these concerns become factors that can lead one to seek out and identify a mentor who meets the needs of particular circumstances and times.

It is important to recognize that, as with all human relationships, mentoring relationships are not perfect. Keeping in mind the following points will help you make informed decisions about when it is wise to terminate a mentoring relationship and seek new ones.

1. Not all mentor/protégé relationships are positive.
2. Most mentoring relationships are either one-on-one or in groups. It is best to gain a variety of mentors to receive diverse information instead of just one person's opinion.
3. If a mentoring relationship begins to get to intense, then it will be harder to make a break when necessary. If a mentoring relationship is more casual, then it makes beginnings and endings easier.
4. It takes time to develop a strong mentoring relationship, which may cause some isolation from other interactions.
5. Often mentors set the tone for the mentoring relationship rather than consulting the protégé.
6. Mentors and protégés may not know information that the other requires at certain times.

Table 3.2

IMPORTANT QUESTIONS FOR PROTÉGÉS TO CONTEMPLATE AT ANY TIME

What criteria will you use to pick your mentor?
How successful do you want this mentoring relationship to be for you?
Do you want the experience to help you to become a mentor in the future?

How do you expect the mentoring relationship to help your career in your chosen field of science, technology, engineering, or mathematics?
What previous advice have you received with regard to women looking for mentors?
What do you expect to be the best advice/experience/knowledge to be gained from your mentor?

What advice would you give women looking for mentors?
What is the best advice/experience/knowledge gained from your mentor?
What mentoring programs do they promote in the workplace/school?

During your experiences, did you feel that you needed a mentor more or less?
Do you still keep in touch with your previous mentors?
Why did you wait until recently to look for a mentor?

What were your initial thoughts about starting a mentoring relationship?
What has this mentoring relationship helped you to learn about yourself?

3.2.1 Points to Ponder

- Identify your most critical concerns at your current career stage.
- Reflect on positive and negative aspects of previous mentoring experiences.
- Ask yourself "what do I need from a mentor now?"

3.3 Techniques and Tools for Starting a Mentoring Relationship

> Do not sell yourself short, and don't assume that anyone is too important or too busy to work to help you out.
>
> Sherry Marts

By tradition and by practice, a mentor has been defined as someone who is more experienced and who provides advice and direction to the protégé. Such a mentor can be from one's own institution, or from one's own field of endeavor, or from an arena into which the protégé wishes to move. In this role, the protégé should expect that a mentor would create opportunity for the protégé, identify ways in which the protégé can train for leadership, and enhance the professional development and career prospects of the protégé.

In this capacity, mentors play key roles in fostering opportunities and in creating environments that can be more hospitable and welcoming for protégés, particularly at the beginning career stages. Such roles include, particularly for young women, ensuring that: (1) research training is available in all STEM fields; (2) careers are not sidetracked when scientists have interrupted careers to fulfill family obligations; (3) outreach efforts exist for young girls and women who have interest in pursuing careers in STEM; (4) professional societies encourage support of career advancement of women scientists; and (5) networks of support exist.

Table 3.3

What Are My Specific Responsibilities as a Protégé?

1. Stick to the agreed upon schedule.
2. Give and receive positive feedback when discussing various issues.
3. Set goals and have a plan to accomplish them.
4. Review goals, outcomes and accomplishments on a consistent basis.
5. Ask questions and be an excellent listener.
6. Seek out resources and opportunities that can accelerate your career.
7. Always maintain a positive attitude.
8. Demonstrate and practice principles learned from your mentoring relationship.

3.3.1 How Do I Pick a Mentor and Start the Process?

First, determine what you would like to learn through your mentoring relationship. For some, this may mean a focus on enhancing your skills and abilities in certain areas such as public speaking, scientific writing, or setting priorities. For others, it might be a focus on understanding the best workplace to pursue your professional interests. Other areas in which you might desire the perspectives and guidance of a mentor could entail work–life balance, life–partner issues, or adapting to unfamiliar cultures or geographies. It is up to you to determine what you seek in a mentor at a given time. Then, you can do research on professionals who have similar interests or who are in a position or a place to provide the perspectives you seek. Once you have identified a potential mentor, simply send an email (or call or write) to introduce yourself and explain why you are contacting them. Of course, if you are participating in networking events at scientific meetings or in other contexts, these also provide many opportunities to make contacts with possible mentors.

3.3.2 What Questions Should I Ask?

A simple way to start the first conversation with a mentor is to ask the following questions. These questions may be used in an ongoing mentoring relationship, or in contexts in which you meet individuals for the first time. These topics work especially well in networking situations and in circumstances where you might initially feel awkward or out of place. It is a rare scientist who will not expound at length on these questions.

1. How did you get started in your particular career?

The many ways in which professionals have reached their current career positions can be very illuminating. It may be as a result of carefully planned steps along the way, or it may have been a pathway around obstacles that generated new directions. Each person's story and choices will provide perspectives that you can assimilate into your own decision-making processes.

2. What do you like and not like about your particular field?

It is a rare professional who likes everything about her/his field of endeavor or particular job track, though it may be hard for some individuals to admit that on occasion. When someone waxes eloquent on all the wonderful things about her career choice, you can "map" her enthusiasm to what intrigues and sounds like a match to your own interests. Hearing from those who can objectively discuss the less desirable (to them) aspects of their career is also extremely useful. No field of endeavor provides a perfect career for anyone. Having a good understanding of the varied career environments and the people who populate those environments can help you make better informed choices and decisions.

3. What are the advantages and disadvantages of your particular career choice?

As with the questions above, a good mentor will provide honest answers that are neither too negative nor too positive in reflecting career choices. Everyone has had to make tradeoffs on career pathways, whether they are likely to admit it or not. Many individuals who are asked this question can describe how their interests, aptitudes, and personality match the career they have chosen. While not everyone may articulate strong self-knowledge about his or her motivations and aspirations, you will be able to acquire information helpful in your own decision-making.

4. What would you do differently if presented with the opportunity to "start over"?

Asking this question can often lead to some very interesting answers and discussions. Some people would not choose to do anything differently, while others would make quite different choices. As long as the response is placed in a context, it can provide additional background for further questions that you might want to ask. Some answers may be solely focused on choice of professional field and other answers may reflect life choices that had to be made because of economic or geographic constraints. There are no right or wrong answers, insofar as the intent is to assure that the protégé can ponder what is really most important to her/himself.

3.3.3 Are There Any "Rules" I Should Follow?

For those people with whom you will have an ongoing protégé–mentor relationship, you will probably want to establish expectations early on. Not only will this make the interactions more productive, but also will serve to minimize possible misunderstandings. First, you will want to get to know each other sufficiently well so that you can mutually set goals to accomplish during the mentoring relationship. Secondly, you should set guidelines for how you would like to proceed, including mode of interaction and topics that may be "off-bounds" during the mentoring experience. These do not have to be rigid, but are a way that both protégé and mentor can articulate realistic expectations about what is to be encompassed within the mentoring framework. Thirdly, it is important to have a regular meeting or contact schedule, at a frequency with which you both agree. Setting the interval can be a reflection of how acute are some of the issues of concern or of the mentoring support needed. And, most importantly, be sure to express a heartfelt "thank you" to your mentor often.

As you embark on mentoring relationships, there are a number of aspects that will not only enhance the mentoring relationship but also help develop other aspects of your career. These are concepts that will work well in the long run in both your career and life: being honest, forthrightly confronting the sometimes painful realities; not trying to redesign the past; and following through, closing the loop, and finishing the project.

3.3.4 Points to Ponder

- Think about your strategies for identifying mentors.
- Quiz yourself as to whether your current mentors are meeting your needs.
- Develop a plan for acquiring new mentor(s).

Table 3.4

What Other Key Advice Should Protégés Consider?

1. It is ok to have more than one mentor to provide a non-biased opinion.
2. Determine what you hope to gain and bring to a mentoring relationship.
3. Make it clear to your mentor what you hope to gain from your mentoring experience.
4. Be able to listen and take criticism to apply to areas that need to be improved.
5. Share future programs, accomplishment, and findings with your mentor.
6. Ask your mentor about possible networking opportunities.
7. Consult your mentor on future endeavors and career planning.
8. Discuss availability with your mentor so that they can provide good service.
9. Remember that your mentor is not an expert in every area but will help you to the best of her/his abilities.
10. Take advantage of all networking opportunities and do not be afraid to talk to people who are not like you, but have similar interests.

Chapter 4
Mentoring Relationships

4.1 What Makes a Relationship Work

> A mentor should not have direct control over your advancement in your organization—the relationship should be 'power neutral.'
>
> Karen Houseknecht

There are a number of elements to a successful mentor–protégé relationship. Each element is a reflection of the temperament, interests, experience, and perspectives of the individuals involved. Not all encounters at all times with a mentor will be uniformly positive. Sometimes our mentor may be distracted by concerns about her/his own life and not able to focus on what we feel we need at that particular moment. Recognizing the myriad aspects of not being perfect human beings, we can be prepared to more productively engage with our mentors. There are some principles which can help us to have successful relationships and interactions with our mentors. Obviously these principles are equally applicable to the mentors themselves.

4.1.1 Be Yourself and Do Well by People in All Your Interactions

This principle is both a call to be truthful and honest in your dealings with others and an ideal to strive for in all aspects of your life and career. A critical element here is trust, both in yourself and in your mentor. To obtain the advice and perspectives that you seek, it is essential that you trust your instincts in facing the uncertainties before you. Knowing the values that are most important to you will help you assess whether a mentor is actually a good match for you. This does not mean that we must share the same value systems as our mentors, but rather that we be able to assess whether our mentors have our best interests at heart.

4.1.2 Never Embarrass Your Mentor or Put Your Mentor in an Awkward Position

We all have moments in our lives and careers that are painful to recall; some may actually involve our current or past mentors. While one should never deliberately compromise one's mentor, inevitable misunderstandings can occur. In that case, a simple "I'm sorry" can go a long way toward maintaining or restoring a healthy mentor–protégé relationship. As with every other aspect of the relationship with a mentor, one can use this as a learning opportunity. In some cases, the storm is weathered and the relationship can continue. In other cases, it may be necessary to move away from this particular mentoring relationship. The perspective of hindsight and the wisdom that comes with lessons learned can help us to develop recovery strategies from mistakes or missteps.

4.1.3 Look for Patterns in Your Life and in Your Career

Working with a mentor can help us acknowledge both positive and negative patterns in the choices that we make. This more objective view, brought by our mentors, can help us identify key influences in and over our lives and careers. To understand the elements of our successes and failures, the reasons for choices we make, and the forces outside our control can position us to make better decisions. For some people, having a different job in a different organization every 3 or 4 years may be an important positive undertaking; for others, developing a career-long commitment to a given workplace may be paramount. The important point here is that, once we understand what makes us more challenged, happy, and fulfilled, the more successful we can be in making critical choices.

4.1.4 Have a Sense of Humor

While there may be a rare person who does not possess even a remote sense of humor, you do not have to be that individual. Having the ability to laugh at yourself can be very cathartic. Moreover, it does not imply that you don't take yourself seriously. Most of us have an innate ability to find humor in most situations, but the critical skill is knowing when and how to deploy that sense of humor. Admittedly, there are some circumstances in which it may be difficult to ascertain whether you should laugh or cry. Mentors, by sharing their experience and observations, can provide many anecdotes of appropriate (and inappropriate) uses of humor. Inappropriateness of setting and of content has derailed more than one career in STEM.

What is essential is that we each find a "comfort zone" for how we can implement a sense of humor that is fully commensurate with our professional demeanor and values.

4.1.5 Recognize that Your Actions, Whether Good or Bad, will Often have Consequences

For anyone embarking upon a career in the STEM fields, "what goes around comes around" is a concept best kept in mind at all times. In fact, this particular principle is a key tenet of acquiring a mature understanding of a lifetime with others. Mentors are a particular resource in helping us grapple with this principle, particularly in the context of subdisciplinary and specialized areas. It is difficult to know where, when, and how often one's path will intersect with others over the trajectory of a career. It does not mean that one submerges one's own interests, personality, or temperament against the hope of future gain. What is more likely, serendipitous encounters often play a more important role than we might realize at the time. Again, mentors can recount their experience and observations as tools to help us see the larger framework in which we can choose to act, rather than to merely react.

4.1.6 Seek the Hidden, Unwritten, and Inside Rules

Mentors who are in our chosen field of endeavor can be particularly valuable in helping us understand the culture and measures of a successful career trajectory in that field. Sometimes this can be as simple as knowing which professional journals are the most prestigious for primary publications or which conferences are the most important to attend. On the other hand, it can be as complex as helping us to understand why some individuals may overly influential in our designated field. At other times, mentors within our own workplaces can help us acquire a more in-depth understanding of workplace dynamics and culture, and of how various people interact. Advice and guidance to protégées is particularly critical for this principle of mentoring. Having mentors who are trustworthy and honest is essential when one is searching for the "rules" that are not specified.

4.1.7 Points to Ponder

- View yourself from your mentor's perspective.
- Look outside your comfort zone for a mentor.
- Recognize that you can also mentor your mentor.

4.2 Mentoring Impact: What Protégés Say

> If a student came to me appearing troubled, I would try to put her in a more positive place by asking what got her excited about science in the first place.
> Donna Vogel

The perspectives of undergraduate and graduate interns in the national office of the Association for Women in Science over the years have provided much excellent insight into key aspects of mentoring. In particular, they have brought to the forefront the issues that are uppermost in the minds of young people just embarking on a STEM pathway. The two sections that follow illuminate aspects that are important to protégés.

4.2.1 What do Protégés Want?

1. We want mentors who encourage and help us to consider ALL opportunities based on our life goals.
2. We want mentors who can help with career as well as life issues (family, friends, and job).
3. We want mentors who will be supportive of academic, career, and personal achievements.
4. We want mentors who are available at any time.
5. We want mentors who will challenge us to think outside the box.
6. We want mentors who will introduce us to other professionals in industry and academia to further our career or knowledge about our specific area of concentration.
7. We want a mentor because mentors give us positive role models in our field to look up to.
8. We want mentors who will be consistent.
9. We want mentors who are like-minded.
10. We want mentors who will help us to become a great mentor ourselves.
11. We want mentors who are honest and trustworthy.

4.2.2 What do Protégés not Want?

1. We do not want a mentor who gives advice based on what interests them and is not necessarily of interest to the protégé.
2. We do not want a mentor who cannot help us consider all options when making a major decision.
3. We do not want a mentor who lacks interest in the areas we choose to study.
4. We do not want a mentor who is not supportive of major accomplishments.
5. We do not want a mentor who does not provide ways to help advance careers in the STEM fields.
6. We do not want a mentor who is not available to provide assistance.
7. We do not want a mentor who will deter us from set life-long goals.
8. We do not want a mentor who will have a negative impact on our life.
9. We do not want a mentor who will not bring out our full potential.
10. We do not want a mentor who is biased toward (or against) women.

Getting the Most out of Your Mentoring Relationships

Lilian Perez was the advocacy and public education intern at the national office of the Association for Women in Science in 2007–2008. With an undergraduate degree in neuroscience, she is currently completing a master's degree program in public health at Emory University in Atlanta, Georgia.

A mentor is someone you feel comfortable approaching – even with personal issues. I find that many times we tend to think of a mentor as strictly a resource for work or school. While they can be great guides for both, I like to look for someone who I can turn to even when something happens in life, as those life events also affect our productivity. A mentor listens and provides advice but also honest criticism. I've had mentors who only praised my work without criticizing where I needed to improve, which is crucial when you are trying to learn and grow. A mentor also needs to be accessible. Communication is the key to having a good relationship. When beginning your search for a mentor, look for someone who you feel can understand your concerns but know that a mentor can be someone with a different background than yours. Sometimes people like to choose mentors who have high credentials, and while that can be helpful, it can also mean that they are very busy people who may not serve you well when you need a minute of their time to talk.

My relationships with my mentors have helped me mature tremendously. They have helped me improve my confidence in myself and in my work as well as taught me to keep pushing through the challenges that I face. I enjoy receiving guidance from others as they can see where I'm heading when I feel like I lack direction. Not all the advice I have received has been good, but with experience I can now step back and steer my own course. In this sense, my mentors have taught me that I have options in life but that only I can decide which path to take.

Seeing other women like me, i.e. Latinas in science, really made a difference when I was pursuing a neuroscience major in college. Seeing these women triumph above the challenges facing women in science really inspired me to keep going when I felt like maybe science just wasn't for me. I felt that if they could do it, so could I. Having a strong network at my institution of under-represented minority women in science also helped when it came to finding support and guidance. My mentors have been excellent role models, they have expressed extraordinary characteristics that I have come to admire and hope to reflect.

Interning at the Association for Women in Science office last year made me realize how amazing maintaining strong relationships with your mentor can be – women mentoring women in some of the most challenging fields out there is an extraordinary sight to see, as many women scientists today are very busy trying to excel in their fields while raising a family that it can become difficult to spare some time to mentor others. I truly applaud all the women scientists mentoring other women in science out there!

Jessye Bemley was a summer intern in 2008 in the national office of the Association for Women in Science and is a senior at North Carolina A&T University majoring in engineering.

As a young girl growing up, I was surrounded by science and technology. In grade school, extracurricular programs, and college I dabbled in a number of different fields and finally settled on industrial engineering. When I was accepted to North Carolina A&T State University, I began to seek out positive female mentors who could help me during the course of my college studies. During my sophomore year, I met my advisor and had no idea she would become one of my greatest mentors. At first, we just met about scheduling and class loads. Then we began to focus on what I wanted to achieve. I mentioned to her that I wanted to participate in the North Carolina Louis Stokes Alliance for Minority Participation Program to do research and that I was searching for a mentor. She agreed to let me work on some research she was conducting on Markov processes. It was really beneficial to start working this early because it prepared me for her class on operations research, which I took the next year. She also nominated me for the Travis Powell Award, college recognition for being helpful, caring and having good interpersonal skills. During our daily meetings I always discuss classes and other issues about life and where I plan to be in the future. She always gives me great advice that is non-biased but straight to the point. I always mention how great a mentor, inspiration, talented, and accomplished person she is to everyone I know because it helps me to continue to strive for excellence. I am happy to know a wonderful person that makes such a positive impact on my life.

Liz Klimas was a summer intern in 2008 in the national office of the Association for Women in Science and is a senior at Hillsdale College (Michigan) majoring in biology.

I never had anyone I would call a true mentor until the summer before my senior year at college. I was staying for 6 weeks during the summer to conduct my senior research. While I spent a fair amount of time with my research advisor, I found myself getting to know another professor during my downtime at the lab. She got me involved with working in the greenhouse and eventually recruited me to help other professors with gardening for some extra money.

I did not really consider labeling her as a mentor until recently. Because she was outside of my research project, I was able to turn to her in discussing this with an outside perspective. She also taught me a lot about botany, which I never would have realized I loved if it weren't for her. I visited her office everyday as part of my routine and we would talk about varying topics from the latest movie we both saw or what we were doing for lunch, to more serious subjects such as my upcoming summer internship with the Association for Women in Science – which was only a few weeks a way – my future career, and occasionally personal events in my life.

The most important characteristic of my relationship with my mentor is my unfailing trust in her. I know I can tell her anything and she won't judge me and she often has great ideas or solutions. I also appreciate her honesty. Though brutally honest at times, our relationship is such that I always respect what she has to say and often come to the conclusion that she is right. My mentor is such a good example without even knowing it. It is easy for me to relate to her because she is relatively young, and I have a tremendous amount of admiration for her. Though only 33 years old she has worked at my college for seven years with a Ph.D. and just recently got offered tenure. It is encouraging to see her, at such a young age, with a Ph.D. and tenureship. She has all this while maintaining a fun-loving and down-to-earth personality.

Oju Ajagbe was a summer intern in 2008 in the national office of the Association for Women in Science and a senior at Pennsylvania State University majoring in life science with a minor in health policy and administration.

I decided on my medical science career path of pediatrics in seventh grade. With both parents already in the medical field – my father is a dentist and my mother is a nurse – I felt that I was set. I had my parents to guide me so no one else's "opinion" mattered. (Unfortunately, at that time, I strongly believed in Mentoring Myth #10: The mentor would always feel s/he knows best, not caring about anything I said or believed.)

Even though there were numerous individuals I could have turned to for advice (and my parents would have approved), I did not feel that I was ok on my own and just seeking the counsel of my parents. When I entered college, there were many (school-established) opportunities to find a mentor – whether in the form of a college faculty member or a fellow student. But I didn't take advantage of any program. One bad experience, during the fall semester of my freshman year, helped turn me off to mentors and mentoring programs – from my university. Being an African-American female, and pursuing a science degree, special support programs were in place to aid me in my undergraduate science career. During the first required meeting, I and my peers had to listen to three African-American faculty members basically state that the majority of the people in that room would switch out of their science majors due to the rigorous nature of the courses and 'it's better to just leave now'. That was not encouraging to me at all. I made sure I did not attend future meetings and I am still a science major.

Sometimes I feel that I may have interpreted their words wrong or was quick to judge instead of going and asking them for clarification. Still, even though this potential avenue of support and advice was not useful for *me*, I should have looked elsewhere instead of writing mentoring off completely. Since I trust my parents, I should have asked them to recommend people they knew could have been great sources of counsel while completing my undergraduate studies. Now as I enter my senior year, my decision to pursue a career in pediatrics is not as steadfast as when I was a freshman. While working with children in a health setting is still my passion, I now have an interest in health and public policy. If I had a mentor, or even mentors, in pediatrics, as well as health and public policy, maybe I would not be wondering what to do after graduation.

4.2.3 *Points to Ponder*

- Mentors are a rich source of perspectives on non-work aspects of our lives.
- Mentors can help us confront any outcomes of our bad decision-making.
- Mentors can be strong role models as well as mentors.

4.3 From Protégé to Mentor: Voices From the Field

In my experience, the major difference is not between PhDs and non-PhDs, but rather between those who, upon graduation at any level, felt that their education was over and those who realized it was only just beginning.

Bob Rydzewski

The variety and context for mentoring is as variable as the desires and interests of protégés themselves. Needs change and the questions that protégés pose often lead, in unexpected ways, to new career directions or perspectives. This chapter provides vignettes of a number of individuals who spoke specifically on the topic of their experience in mentoring, in responding to a selection of questions such as those below.

4.3.1 *Questions Posed to Mentors*

What do you like most about your mentoring experience?
What has this experience taught you about mentoring?
What do you expect of your protégés?

Why do you think it is important to have more female mentors in the STEM fields?
What feedback do you receive from your protégés?
How successful are your mentoring relationships?
What advice would you give other mentors?
What was the best advice given to you?
How do you think e-mentoring differs from other methods of mentoring?
What advice do you have for mentors to prepare for mentoring relationships?
Did you have a mentor inspire you to consider mentoring relationships?
What mentoring style would you recommend?
What would you consider an effective mentoring relationship?
What is your take on mentoring in general?

4.3.2 Many Mentors, Many Perspectives

Dr. Andrea Stith is a guest researcher at Humboldt University of Berlin under the auspices of the German Chancellor Program of the Alexander von Humboldt Foundation in Germany. She is working on a comparative study of how Germany and the U.S. fund graduate science education, design research training programs, and promote opportunities for early career scientists.

> Before I went to Germany as a guest researcher, I needed to find a mentor as part of the Humboldt Foundation's German Chancellor Fellowship. The program is under the patronage of the German Chancellor and also incorporates an intensive language course in Germany, a four-week introductory seminar in Bonn and Berlin, a study trip around Germany and a final meeting in Berlin. These activities provide additional insights into the social, cultural, economic, and political life of Germany.
>
> Unfortunately, just as I was beginning my work in Berlin, my mentor got transferred to a position in the U.S. When looking for my initial mentor, I wanted to find someone in a complementary field to mine. My new mentor is also helpful, but our study topics do not align very closely, so I pretty much rely on the kindness of strangers for guidance. But, as a result of my current mentoring experiences, I have learned that I am tangibly more independent then I was in graduate school. I no longer depend on a single mentor. I look for a broad range of advice and expertise. When looking for mentors, don't be afraid to look for more than one. Not everyone will give you the best advice on everything. Don't waste people's time but always be open to hearing them out. And, show your appreciation!
>
> Also consider peer mentoring. During my college years, at the University of Delaware there was an excellent mentoring program for minorities. I started in the program during the summer before my freshman year of undergrad and took two classes. There were 20 in the program who lived together. We formed a 'miniclass', and our friendships have lasted throughout the years. This program allowed me to move from protégé to tutor and helped me to learn that peer bonds are important. All of us need reflect on how we can provide support to one another. We are all role models with the ability to foster and encourage success and growth in others.

4.3.3 Never Too Busy to Help

Dr. Sherry Marts most recently was the Vice President of Scientific Affairs at the Society for Women's Health Research. She directed several scientific programs such as the series on Sex-Based Biology and the "Some Things Only a Woman Can Do" public education campaign on clinical trials. She was also a scientific spokesperson for the society.

In 1986, I received my PhD from Duke University. As a graduate student, I enjoyed studying science but quickly realized I did not really love laboratory research. I was active in reviving the graduate and professional student government at Duke and I got involved in local politics, including working as a campaign volunteer for a state legislator. One thing I like about my volunteer work was convincing people to work on an issue while letting them think it was their idea. My advisor at the time was supportive of my exploring other careers but was not very knowledgeable about careers outside of academia.

I turned to the Career Center at Duke. I was clear that I did not want to work in laboratory research. The Career Center had never before counseled a science PhD candidate who did not want a career in academic or industry research. The Career Center Director decided to take on the challenge of helping me to find the job I wanted. First, we began by crafting a resume that deemphasized research and highlighted transferable skills from my research experience and my volunteer work. I also did a lot of informational interviews, one of which led to a job offer with a public health organization. However, just as I was poised to join them the company changed its focus to a topic that I was uninterested in pursuing. I spent hours in the library tracking down potential employers who, I thought, might have the need for a trained scientist with skills at organizing people, planning events, writing, and public speaking (and who were in places I wanted to live and worked in areas that I found interesting, of course). I sent out resumes and cover letters that described my ideal job to all the places and organized interviews and meetings. One company I wrote called and wanted to know how I knew about the job before they had even advertised it. They ended up hiring me into my first non-laboratory position.

Throughout this process I relied on a wonderful book, *What Color is Your Parachute?* by Richard Nelson Bolles. I really love this book because it is realistic and is based on what works for creating the job you want rather than on "conventional wisdom" about job hunting. The exercises in the book are great for figuring out what you like to do and then crafting that into a job description. By knowing this information you can go into an interview with a dash of confidence because you know what you want and, equally important, what you do not want. So often when job hunting, the temptation is to accept the first thing you are offered, even if the job falls short of what you really want. Speaking from experience, that is a big mistake. It is hard to do your best at a job when you are unhappy. So it is worth taking the time to consider what you like to do and try to come up with a job that fits.

When considering a mentoring relationship, find someone who is honest. It may help to look outside of your workplace or outside of your department because your immediate co-workers may not always be comfortable giving the advice or guidance that you need or someone to bounce ideas off to make sure they make sense. Do not be shy when gathering information and ideas about jobs and careers; people love to talk about what they do when they are passionate about it. Do not sell yourself short, and don't assume that anyone is too important or too busy to work help you out. Just ask! And when the answer is no, follow up with a question "Is there anyone you would recommend that I talk to about ___?" Remember that graduate study and laboratory research provide

4.3.4 Finding a Positive Place

Dr. Donna Vogel is currently the Director of the Professional Development Office at Johns Hopkins Medical Institutions. She works with graduate students, postdoctoral fellows, and early-career faculty in the Schools of Medicine, Nursing and Public Health. She loves to interact with students and help them to make the best decisions respective to their situation.

When I joined the National Institutes of Health (NIH) in 1980, career networking was not recognized as the key to finding your next job. There was very little guidance about careers particularly with regard to anything other than academic medicine or bench research. When I was looking to make a career move out of research, a colleague suggested I "talk to Toni Novello." Dr. Antonia Novello was to become the first Hispanic female Surgeon General but at that time she was Deputy Director of my Institute. In that position, she ran NIH's Extramural Program. Dr. Novello looked over my curriculum vitae, gave me some general advice, and then at the end, said that she "might have something" for me. In fact, she knew of a position – that hadn't even been advertised yet – as director of the Reproductive Medicine grant program. In a few days, I got a call from the Branch Chief and was hired shortly after. I had that job for about 13 years with increasing responsibilities along the way.

More than ever, it is vital for women to seek out advice and build mentoring relationships throughout their career. My advice is to start looking at the relevant professional societies. Student or trainee memberships are inexpensive, include the online journals, provide targeted job announcements, and often have career development information and mentoring opportunities. If the society has a trainee group or trainee affairs committee, it will include senior scientists who have made the choice to work with students and fellows: they can be useful allies. When you attend seminars and conferences, you may identify future mentors. If you are looking at a new institution, see who has won teaching or mentoring awards. Whenever considering someone as an advisor, see whether he or she has a good track record: where have students and fellows gone, what are they doing now? Talk to the trainees, past and present, to learn whether they received effective mentoring. You may need more than one mentor, and that's fine, especially if your research mentor is new to the faculty. A more senior "career mentor" can help you with aspects of career growth beyond your actual project.

You should also understand that mentoring relationships can either be very effective or they can be inadequate. Successful mentoring programs require institutional support, integrated records and fiscal management, and support for cultural change in the form of accountability. That is, evaluation of investigators must include their mentoring ability. Many supervisors of postdoctoral fellows still are looking for hands to do research, not committing to mentor the trainee. For many years, NIH acceded to this mentality even in its training and career grant applications. The training plan or career development plan section was simply the didactic coursework and technical training the student or trainee would receive. Only recently have the funding agencies awakened to the need for what we call career development – the professional skills one needs to become an independent scientist.

As a mentor I expect my protégés to have some idea of what questions they want to ask – what will be useful for me to tell them; to take what I have offered, to follow up on introductions and actually use the advice given; and to keep the relationship going by giving me feedback so I can tell if my advice is useful and satisfactory. If a student came to me appearing troubled, I would try to put her in a more positive place by asking what got her excited about science in the first place.

4.3.5 E-Mentoring to Attain Workplace Success

Dr. Joanne Kamens is the Director of Discovery at RXi Pharmaceuticals, a discovery-stage biopharmaceutical company pursuing the development and commercialization of proprietary therapeutics based on RNA interference (RNAi) for the treatment of human diseases. She was co-founder of the AWIS Boston Chapter, now one of the largest in the country.

I was inspired to become a mentor by many wonderful people and experiences throughout my career. My parents, for example, were gender blind and my advisor was really helpful in supporting my goals. Because of these strong role models early on, it was important for me to give back to the sciences, not only through my lab work, but through social and professional networking as well.

When I first joined the corporate workforce at a pharmaceutical company, there were only a handful of women. As the company grew, many of those women left for various reasons, both personal and professional. It soon dawned on me that I was in a striking minority. One day I was sitting around in my fifth meeting of the week, and I realized I hadn't seen a woman in five days! So another female colleague and I formed a women's group, organizing lunches and other mentoring and community-building events in order to keep women in the labs. That is why I am a big proponent of mentoring. It is the key to a successful career.

There are many different ways to build a mentoring relationship. I have found that e-mentoring through MentorNet allows me to connect with a protégé from anywhere in the world and work on my time schedule so I do not feel overwhelmed. I have high expectations and leave the relationship in the lap of the protégé. They should be clear about what they want to get out of the experience and take a leadership role by setting up meetings, returning phone calls, and sending frequent e-mails.

To have an effective mentoring relationship a mentor and protégé must take time to develop a good rapport and protégés need to be responsible about keeping up their end of the "conversation." If not, the relationship is on a sure path to failure. And, protégés looking for mentors need to be aggressive. Look for men as well as women. All you have to do is ask a prospective candidate – all they can say is no and they will be flattered you asked. Another great mentoring relationship is through a Mentoring Circle or among your peers. Peers are often easier to talk to about issues and problems and a lot of people benefit from these types of relationships.

4.3.6 Tapping into the Pipeline

Dr. Patrice O. Yarbough recently served as the Interim Executive Director of the Office of Strategic Research Collaborations at the University of Texas Medical Branch at Galveston in Texas. An ardent advocate for underrepresented minorities in STEM, she is the Chair of the Association for Women in Science Diversity Task Force.

As a National Science Foundation Minority Fellow in 1985, I experienced a total lack of role models and peer-mentors. Since that time, women from minority groups continue to be overwhelmingly under-represented in science and engineering. They are only 3% of all science and engineering faculty in four-year colleges and universities. Furthermore, African American, Latina, and Native American women continue to be under-represented in STEM careers requiring a doctorate. And all minorities, especially women of color, come into STEM careers with a smaller network than average.

One proven strategy for assisting women in STEM fields is the development of a variety of mentoring relationships including peer mentors, career mentors, and personal (work/life balance) mentors. Although many people consider mentoring as only the relationship between a senior, experienced person helping a junior person, the benefits of peer-mentoring are often overlooked. Peer- mentoring relationships are formed during graduate and post doctoral studies and allow the opportunity to interact with people that can identify with similar professional experiences. Peer mentoring is particularly effective for women as it takes the power out of the relationship.

First and foremost, you need to take responsibility for directing your own educational and career development. Talented minorities have many opportunities. This fact should work to your advantage. Seize these opportunities. Look around for unexpected and unconventional opportunities in professional and governmental organizations like internships and fellowships. Join minority organizations as you will find that networking is an important defense against feeling isolated and a useful offense for furthering your career. Take advantage of opportunities to learn how others have solved problems that you are facing and to develop strategies to effect change useful for your community. Learn from your peers which institutions, programs, and laboratories nurture people of all races and ethnicities.

And, no matter how junior you may feel, always remember that you are a role model and can mentor your peers and those who are junior to you. Take every opportunity to share what you have learned with those who are following in your footsteps. Support and be an advocate for others.

4.3.7 Mentoring Is Colorblind

Dr. David Burgess is a Professor of Biology at Boston College. A past-president of the Society for the Advancement of Chicanos and Native Americans in Science, his grandmother was a Cherokee medicine woman and he has been a tireless advocate for underrepresented minorities.

My father was a mentor, a coach, a high school teacher, and principal who worked with people of all backgrounds, especially with minority groups. When he retired, he was honored by the community for his work across cultures and socioeconomic backgrounds. He was able to handle tough situations that required him to be respectful of all people. His commitment to serve others left a lasting imprint on my life.

As an advocate for underrepresented minorities in science and technology, I have learned that minorities who are most successful and advance the furthest share one characteristic – a strong network of mentors. It is important to have more than one mentor to bring new perspectives and share experiences across disciplines, positions, and racial and socioeconomic dimensions. Look forward mentors in communities outside your own.

Getting the Most out of Your Mentoring Relationships

When starting a mentoring relationship there are a few things I expect of my protégés. I think it is important for the protégé to follow through on communication, to keep appointments, to be honest with how they are doing, and to feel free to bring up any issues that might arise. Communication is definitely the most essential part to the relationship.

I, in turn, have also learned a lot from my mentoring relationships. A mentor must be honored that someone is seeking their advice. A mentor must be colorblind – being of the same race or gender as your protégé does not necessarily enhance your mentoring relationship. Mentors should encourage their protégés to have relationships that go beyond academic and work-related issues to build broad, diverse networks that include genuine and long-term relationships.

And, you are never too old to be or need a mentor.

4.3.8 Power Neutral Mentoring

Dr. Karen Houseknecht is the Vice President of Global Research & Development at ASDI, Inc. She is responsible developing and executing the global research strategy for the company and for initiating and developing collaborations with biotech, academic and non-governmental organization partners.

I cannot emphasize enough how important it is to have a mentor, or better yet, a portfolio of mentors to help you navigate and succeed in today's corporate environment. An important thing to remember about corporate culture is that business is all about relationships. From working on teams to solve problems, to partnering with external clients and corporate partners, relationships are the underpinning of success in business. That is also true in terms of career development in today's very competitive global technology environment. There is a lot of highly trained, highly motivated scientific talent competing for the best positions and promotions. Thus, often the final decision on a promotion, or even the heads-up about a new leadership opportunity (especially at high levels), is influenced by relationships. This can include mentoring relationships as well as career networks created both internal and external to the company. Mentoring relationships help you to advance in your career by teaching you important skills such as negotiation and navigating within the culture that is specific to your company or organization – things they never taught you in graduate school. More importantly, internal and cross-company mentoring relationships help you achieve a variety of goals and objectives that are directly tied to your organization's business strategy.

Throughout my career, I have had very few women mentors primarily because there is such a paucity of senior women in scientific roles – both in the academic and corporate sectors. Thus, most of my mentors have been men, and they have been a great help to me during various career transitions. I want to encourage women not to shy away from choosing strong male mentors for their personal mentoring team. Men in strategic positions can provide critical feedback on how you are perceived in the workplace and will often help you develop the skills you need to move up the corporate ladder. Having a mentoring team, ideally a diverse set of women and men, will give you the most balanced and valuable feedback and ensures that you need not depend on one person to help guide you. It is important to remember that mentoring relationships are like all human relationships – they require time, energy and management in order to flourish. It is up to you to manage your mentoring relationships. Clearly define exactly what you want from your mentor. Own the relationship. Invest time in the relationship. Be selfish and devote time to your own professional development and needs.

One very important thing to remember: A mentor should not have direct control over your advancement in your organization – relationships should be "power neutral" in order to be of most value. Your mentors should not include your boss or other members in your direct reporting line in the company. You need to be able to speak openly and freely to your mentor. Remember that your supervisor and others in your reporting line are responsible for delivering on the goals of the organization and those may not always be aligned with your best interests from a career development perspective. An alternative to mentoring is coaching. Recently, I found a professional executive coach who works with executives in technology companies. I chose to explore coaching vs. mentoring at this point of my career because as I have advanced higher in the corporate ranks, I have fewer and fewer people within my company I can go to for mentoring. Additionally, with my increased travel and workload, it is becoming harder and harder to find time to build and maintain my mentoring networks. While coaching and mentoring have many similar characteristics, a coach is an expert on people and personal development. Typically a skilled specialist regarding a certain topic, competency, or industry, a coach's role is to provide structure, foundation, and support so people can begin to self-generate the results they want on their own. Coaching is a process of inquiry, relying on the use of well crafted questions, rather than continually sharing the answer to get people to sharpen their own problem solving skills. In coaching, the relationship is objective, and the focus is not only on what the person needs to do to become more successful but also who the person is and how she thinks. A coach works on the whole person and is multidimensional, rather than focusing only on what the person is already doing.

So explore your options – mentoring and coaching can provide skill building and support systems that will help ensure your sustained business success.

4.3.9 Age Doesn't Matter

Dr. Cathy Middlecamp has a joint appointment in Chemistry and Integrated Liberal Studies (ILS) at the University of Wisconsin-Madison. She teaches "Chemistry in Context" for non-science majors as well as an upper level course in ILS, "The Radium Girls and the Firecracker Boys." She is the winner of the 2006 Camille and Henry Dreyfus Foundation Award for Encouraging Women in Careers in the Chemical Sciences.

As the years roll by, I find myself talking with people 30 to 40 years younger than I am. They have so much energy! And they know about things that are entirely off my radar screen. The stereotypes about the "subject matter expert" have been shattered, as we all have the potential to be a subject matter expert. One's age does not necessarily indicate who is more knowledgeable about what. Mentoring is mutual in the sense that those in different generations truly can learn from each other.

I was inspired to serve as a mentor when I noted how many people were just stopping by my office, wanting to talk. I leave my office door open and do my best to make everyone who stops by feel welcome. Students may come to my office to ask questions, but I do not consider this a mentoring relationship. Rather, it is those who come back over the months and years with whom a mentoring relationship may grow. Most of these students never formally ask me to serve as their mentor; there seems no need to.

Women who are looking for a mentor should not limit themselves to looking for a woman or a person that seems in some way to be like them. People who are quite different from you can make great mentors, and at first you may not recognize them as such. A mentor can be young or old, male or female. Mentors are not necessarily your friends. And try not to rely on a single individual for insights, guidance, and wise words of advice. Rather, look for several people who each can play a slightly different role and learn from the strengths of each one.

The ways in which we can relate to people are changing. Technology is connecting younger people in new ways that some of us older folks never dreamed of. Used well and creatively, technology can bridge the gap between younger and older generations. And of course used badly it can separate them. In mentoring, technology certainly has great potential.

4.3.10 Points to Ponder

- The perspective of mentors provides insights into our own mentoring needs.
- The best mentors have had both good and bad mentoring experiences themselves.
- Being mentored is a life-long endeavor.

Chapter 5
Changing Dynamics, Changing Needs

5.1 Mentoring for Under-represented Groups

If you had a positive experience, go find someone else to bring up the pipeline.

Patrice Yarbough

The numbers of women and under-represented minorities in science, technology, engineering, and mathematics (STEM) disciplines have greatly improved over the past several decades, but institutional efforts to recruit, train, retain, and promote their participation in these fields are still inadequate. Being successfully mentored is particularly important for these groups. Individuals and institutions must step forward to take a leadership role in the advancement of diversity in the STEM disciplines. To include persons of color in the scientific workplace is a moral imperative, as it will increase the variety of perspectives, life experiences, and approaches, in turn, serving to enhance science. Mentoring is of particular importance to the individual person of color in parallel with institutional efforts. A number of issues germane to these particular mentoring needs are discussed below, which embody aspects with which protégés should become familiar.

Although the number of under-represented minority scientists and engineers is increasing, the academic environment has a continuing need for programs directed toward recruiting and training more under-represented minorities in STEM careers. Hence, academic institutions have undertaken initiatives that promote the advancement of such students; develop new programs that offer such students opportunities to engage in activities related to scientific fields, such as research; regularly evaluate the progress of new and existing programs and determine areas where improvement is needed, using surveys to determine the opinions of students, faculty, and administration; and increase the representation of under-represented minorities in faculty positions to facilitate mentoring relationships between students and STEM faculty members.

Although the "urban myth" is that the number of under-represented minorities qualified for faculty positions in science and engineering fields is exceedingly small, studies supported by the National Science Foundation (and other research agencies) show that the proportion of under-represented minorities awarded degrees in these disciplines has risen markedly in the past 30 years. The pool of African Americans and Hispanics with doctorates in science and engineering is therefore considerably larger than generally assumed. Consistent with the increase in native-born minorities in the doctoral-level workforce in recent years, there is a higher proportion of such individuals among younger workers than older workers. A major question is whether the increasing proportions are reflected in the faculties of colleges and universities, where under-represented minorities serve as important role models for future generations.

Because academic institutions are the gateways and pathways that all professionals in STEM must traverse, it is critical that outmoded ways of thinking be eliminated. Hence, tools and techniques that some academic institutions have implemented include workshops that address diversity and cultural awareness issues for academic administrators and faculty; quality control in diversity training programs and follow-up to determine whether action has been taken and progress made as a result of these programs; professional development for department chairs and deans to inform their leadership decisions regarding departmental climate and evaluation issues; developing goals for the hiring and retention of under-represented minorities and women; and holding chairs accountable for providing equitable start-up packages for newly hired faculty.

Protégés should hold their mentors accountable, in assuring that they have full opportunity to achieve and contribute to scientific disciplines. At the undergraduate level, under-represented minority women are slightly more likely than white women to major in scientific and engineering fields, while at the doctoral level under-represented minority women are slightly less likely than white women to achieve a Ph.D. degree in STEM. Nonetheless, in 2004 and 2005 the number of Hispanic, African-American, and Native American/Alaskan women earning doctorates in science, engineering, and math totaled 1640, representing almost 12 percent of all Ph.D. degrees awarded to women in these 2 years. These numbers suggest that these women should be a good source of future United States scientific leaders. However, science departments of research universities currently show fewer such women among assistant professors than would be expected by their representation among Ph.D. recipients. The under-representation of minority women scientists in research universities shows a pattern that is more similar to that seen with all women scientists than to the pattern seen with minority men among STEM faculty. The nuances of these differences, and the reasons that may lurk behind them, will be critical factors for young scientists to explore in a healthy protégé–mentor relationships.

Exacerbating the lack of women of color as scientist role models for women students is a pattern in which the few successful under-represented minority women are rapidly moved into non-faculty and non-science jobs. Administrators in universities, industry, government agencies, and non-profit organizations who wish to demonstrate diversity are attracted to the idea of getting "two-for one". Consequently, the few women of color who are hired for faculty positions are often asked to assume administrative responsibility for both gender and racial/ethnic diversity, while they are still climbing the tenure ladder. If this occurred after they had achieved the rank of full professor and had made a mark within their scientific community, such recognition might be advantageous. However, asking them to carry the total diversity burden at their universities while still relatively junior not only removes them from the scientific community but, in most cases, undermines their scientific careers because of insufficient time to focus on teaching and research. Most of these individuals also feel a strong commitment to mentoring because of the challenges they themselves faced in moving up in their career. Protégés should recognize, and be sensitive to, the complexity of demands and responsibility often placed on their mentors who come from groups historically under-represented in scientific fields (adapted from Suiter, 2006).

5.1.1 MentorNet – The E-Mentoring Network Focused on Diversity in Engineering and Science

Headquartered in Silicon Valley, MentorNet (like the Association for Women in Science) is a non-profit organization that links students and professionals in scientific and technical fields across the globe. Its mission is to further the progress of women and others under-represented in scientific and technical fields through the use of a dynamic, technology-supported mentoring network. Through collaborative partnerships with a variety of corporations, institutions of higher education, government laboratories, and professional societies, MentorNet programs have reached more than 16,000 students and a like number of professionals since 1998. Relying on email and the internet to support one-on-one mentoring relationships, undergraduate and graduate students, postdoctoral fellows, and early career faculty at universities are placed in contact with professionals in industry, government, and higher education as their mentors.

To supplement and complement their other mentored activities, protégés can use the resources of MentorNet (www.MentorNet.org). Protégés can create and sustain mentoring relationships that assist them through their academic preparation and early professional development, particularly significant for fields in which the numbers of women and people of color are still small. A formal mentoring program matches protégés from over 100 participating campuses and affiliated partners to mentoring counterparts in industry, government laboratories, or higher education. Since 1998, MentorNet has matched more than 16,000 pairs of mentors and protégés. MentorNet also provides an array of resources such as discussion boards, a resume database, and links to relevant publications and websites. Topics explored include categories such as diversity issues, academic career development, career options in biology, and professional licensure. There is also an up-to-date recommended reading list on topics including work–life balance and self image.

Young scientists can look to MentorNet for advice and support that either complements or is not available at their current institution. For example, MentorNet protégés within academia can be paired with professionals in industry, who aid them in planning a roadmap to the private sector.

MentorNet also hopes that by being mostly internet-based, it creates accessibility to mentors for geographically isolated protégés as well as those who are of historically under-represented groups in science.

Section 5.2 discusses some additional creative ways in which the new internet technologies can be deployed so that protégés can acquire more effective mentoring. These technologies hold great potential for protégés who are geographically or culturally isolated from their communities and homes. The ability to forge supportive online communities for under-represented minorities has to date been quite underutilized.

5.1.2 Points to Ponder

- Recognize that under-represented minorities in STEM may face very complex career issues.
- Seek out diverse points of view as part of your own growth.
- Do your part to foster diversity and to fight prejudice.

5.2 Mentoring in Cyberspace

> I always wondered why somebody didn't do something about that. Then I realized I was somebody.
>
> Lily Tomlin.

With the dramatic changes that have occurred in the way information can be transmitted, the opportunity to use all modalities available for being mentored should be exploited. For example, membership in professional organizations often includes the option to subscribe to listservs that engender much discussion within a particular career area. Participation can be a springboard that protégés can use to develop mentoring friendships offline. The central message of this "how to" book has been to assist in all career stages, given that most of us realize that we need mentoring throughout our lives. The localized and in-person opportunities to be mentored should not go untapped. However, there is an ever increasing variety of electronic resources that can supplement existing mentoring relationships in one's professional life or that can open doors to developing new ones. Some compelling examples of the emerging and ever changing use of technologies are provided in this chapter.

Most would agree that the information technology age has created new communication challenges in a scientist's life, particularly in the area of balancing home versus work. But the silver lining in this cloud is that such advances have also allowed scientists to take advantage of new mentoring models that would not have existed 10 years ago. Private chat rooms, email, VoIP (voice over internet protocol), video conferencing, and an explosive volume of Web 2.0 tools provide new methods for the mentor and protégé to connect and therefore give scientists across the globe greater access to mentoring opportunities.

5.2.1 Mentoring in the New Era of Social Media (Web 2.0)

The latest technology on the scene is Web 2.0, or social media, which allows users to communicate with others through various means. Wikipedia refers to Web 2.0 as "enhancing creativity, allowing information sharing and collaboration among users through the world wide web as well as websites." You may ask yourself the question "how can this technology be used for mentoring purposes or by networking to keep up with individuals who can help propel my career?" Currently, there are several people who are dedicated to increasing awareness of how to use the different Web 2.0 technologies in order to reach out to the masses as well as create better bonds between peers and colleagues.

Web 2.0 was coined in 2004 at a Media Web 2.0 conference hosted by Tim O'Reilly. He defines Web 2.0 as "business embracing the web as a platform and using its strengths." From this conference, O'Reilly and his partner, Battelle, came up with four levels to describe how this technology can be used:

Level 3 – Represents most Web 2.0 programs that already exist on the internet. The more people use them, the more effective the applications.
Level 2 – Represents applications that can work offline, but would gain some advantages if brought online.
Level 1– Represents applications that can work offline but can gain different features if put online.
Level 0 – Represents applications that can work well online and offline.

The great thing about Web 2.0 is that you can retrieve, update, and add information on a consistent basis. Four examples are briefly described below, for which the potential is just beginning to emerge.

5.2.1.1 Blogging

Blogging is a way to publish stories on different issues or topics. Blogs allow organizations, companies, and individuals to have a voice about certain issues. These posts, in turn, get other people to voice their opinions. Blogging would be useful in a group mentoring setting – everyone would get a chance to state their views on different issues and a discussion would be facilitated by one or two mentors. The protégées can write their stories or issues in one blog. Then, mentors can post comments that are beneficial and helpful to the protégées.

A mentor can also post a blog on a relevant topic that their protégés can view as soon as they check the blogspot. Blogs can remain on the internet for as long as you want. If it is accessible, the information could be available to a limitless number of interested parties.

5.2.1.2 Tagging and Social Bookmarking

These two technologies allow a person to tag keywords of different digital objects to describe them. Social bookmarking allows users to create a list of favorite websites and share them with other users in a social system. Many mentors can provide information for their protégées about how to prepare for and even excel throughout their careers. Mentors can create their personalized social bookmarking systems, which allows their specific protégées to get information related to career paths, higher education, issues with jobs, conferences, and seminars.

5.2.1.3 Audio Blogging/Podcasts

Portable media players, such as the Apple iPod, are the "must have" items of the Web 2.0 savvy generation. From this device came two new technologies: audio blogging and podcasting. Audio blogging is similar to a regular blog, except that it can be put into an audio format for others to hear. A podcast can consist of interviews, talks, or lectures; they can be played on a desktop or on any handheld MP3 device. If you do not necessarily have time to talk to your protégée or mentor everyday, you can create a series of podcasts or audio blogs related to the topics or areas for guidance. These technologies can also be used to provide generic mentoring help for anyone from any background. If you have never met your mentor or protégée, using video podcasting allows you to see and speak with one another.

5.2.1.4 Social Networking

Social networking allows users to create networks of people who are like-minded, want to share similar content, and create ways to meet new people. These particular websites make it easier for protégés to identify those individuals who may be pursuing careers they consider interesting. The more information that a potential mentor makes available, the easier it is for a protégé to consider you for a mentoring relationship. On these networks, you can look at different profiles and befriend people, using your network specifically toward being mentored or finding mentors. The more information provided about you, the easier it is for a candidate to consider you for a mentoring relationship. Although many people are concerned with knowing everyone as their 'friend' online, you never know when you will come across a person who can help with your career, job, mentoring, or provide simple advice.

Since the beginning of the Social Media era, many people are skeptical about whether or not they should use the different programs and software that lead others into your every move. First, you can provide sites that allow your members or workers to stay or get connected to different people. Some sites to make this connection possible are Yahoogroups, Facebook, and LinkedIn. Personal interaction allows you to talk to people face to face and understand how to use the technology and use it efficiently. Since the beginning of Web 2.0, many people have created blogs, forums, bulletin boards, and other links of information that can be helpful for a project or business and used as a resource. To motivate or promote your team, you can use podcasts, videos, or webcasts that highlight major accomplishments to make others aware of what your organization, school, company, or church has done. Even though you are considered a friend on different social networks, creating a mentoring relationship should not be taken lightly but with a serious attitude. Having so much technology and connectivity makes it easier for you to talk to anyone in any place at any time.

5.2.2 Using Web 2.0 Technology to Empower Specific Groups

James Neusom, the executive director and publisher of InterServe Networks/ City Light Software Inc., has numerous networks on various social media such as Myspace, Facebook, and Blackplanet. Periodically he blogs to keep the African-American community updated on various issues concerning the sciences, engineering, and computing. He also discusses Web 2.0 information through blogs so that people become aware of the many ways in which this technology can be deployed. One specific organization that has used it very effectively is the Black Data Processing Associates (BDPA) science program for youth. Wayne Hicks, currently the Educational Program Director for BDPA and Director of his very own company "Hicks Enterprises", has the opportunity to talk and work with many young people. This allows him to stay current on uses of the various technologies, especially Web 2.0. He has been the initiator of several pages on different sites such as Facebook, Twitter, Myspace, and LinkedIn for BDPA participants to connect with one another. Since he started these pages, more people have been introduced to Web 2.0 and realized that these new tools can be used to create a bigger network of people that may not know what you, your school, company, organization, or business is all about.

Shireen Mitchell is the director of a non-profit organization called Digital Sistas. She uses all types of Web 2.0 media, which helps her receive and send information from users. She writes blogs on various women's issues and even has a channel on You Tube which presents different videos about how girls and women are portrayed in the media in respect to science, technology, engineering, and mathematics. Using these different tools and having them connected to the various websites gives her a chance to create a large network of people who have similar interests.

5.2.3 Science Careers Forum

The American Association for the Advancement of Science (AAAS) has a highly active online forum on science careers. Unlike other science career forums on the internet that quickly become infected with overly political and argumentative discussion threads, this forum attains and maintains a higher professional tone including the use of professional language, respect for all posters even when there are disagreements, and the eschewal of non-productive political diatribes. The result is a high quality discussion format that draws professionals from nearly every arm of academic and industrial science and engineering. There is a wide variety of topics including career transitions, graduate school life, work–life balance, tenure track aspirations, interacting with human resources representatives, and movement within the industry echelons. Participants may post anonymously and choose to make their email address accessible to other posters. The discussion threads themselves range from highly pointed questions to philosophical ponderings about the state of science in the United States. (See http://scforum.aaas.org/.)

5.2.4 Ph.D. Career Clinic

This web site was developed to focus on "help[ing] Ph.D. professionals leverage their doctorate training beyond academia." Excellent viewpoint pieces are posted on all aspects of career development, with readers encouraged to leave thoughts and feedback on the posts. Topics explore areas beyond the typical 'Networking 101' and 'Finding Your First Industry Job' articles located on other websites and publications. Cross links to other career development resources are also a feature of the site. (See http://www.phdcareerclinic.com/).

5.2.5 Points to Ponder

- Experiment with new modes of forming networks of support.
- Form your own peer-mentoring group in an area of common interest.
- Remain current on emerging capabilities that foster new modes of mentoring.

5.3 Life-long Mentoring

> You can't be too old to be or to need a mentor.
>
> David Burgess

From one's undergraduate years, through graduate school, postdoctoral and/or early career positions, to mid and late career stages, rich and abundant opportunities exist for mentoring relationships. The relationships with mentors will change, whether one acquires new mentors at various stages or a mentor remains a mentor through decades of interactions. Both mentors and protégés can and should benefit from a mentoring relationship, but that is not always or necessarily the case. At whatever life stage, mentoring relationships should always be confidential and based on mutual trust, respect, and good communication. Institutional climate and initiatives can also dramatically affect mentoring relationships and their outcomes for protégés.

5.3.1 At the Student and Trainee Level

Mentors can assume a number of roles and can occupy a variety of positions. Such individuals can be college professors, the Ph.D. advisor, department chairs, fellow students, members of the dissertation committee, the postdoctoral supervisor and other senior researchers, or even professionals and professors in other fields. At this stage, protégés should have mentors who work to increase their visibility at institutional site visits by funders, at laboratory meetings, at research planning retreats, and at professional meetings.

Developing critical communication skills is an important facet of career preparation training. A conscientious mentor works with the protégé to enhance her/his oral and written presentation skills. Reviewing manuscripts is a critical aspect of a professional STEM career and protégés should expect their mentors to provide appropriate opportunities for them in this regard. Protégés need mentors who can illuminate the many career options and pathways available for a career in the area in which the protégé is becoming trained. This may often require the protégé to seek individuals outside the institution for a broader perspective. Mentors perform an invaluable role in setting expectations for the protégé and in evaluating the protégé's progress.

Some fields of endeavor require completion of extremely competitive and demanding graduate programs, and students are asked to consider these facts and always examine a broad range of options. Once these decisions are made, students then must have a back-up plan in case either their field of pursuit or even the scope of commitment must change. What better topics to probe with mentors? Dr. Emil Thomas Chuck, health professions advisor at George Mason University in Fairfax, Virginia has noted: "The absence of an alternative career plan is often a sign of a lack of maturity as well as a failure in recognizing the competitiveness of graduate programs.... I do make the point in my advising to ask students to imagine their lives in 10 years and acknowledge that many scenarios can unfold both personally and professionally. I stress the importance of job shadowing and witnessing how trainees are treated. Such experiences can inform the student of unexpected pressures they will face in their chosen profession, and empower them with the knowledge to decide whether it is a good fit for their lifestyle or if a change is in order" (Horvath, 2007c).

Critically important is the selection of a laboratory and a graduate research advisor that suit the personality of the student. As well, it is essential that mutual respect and good lines of communication be established early in the student/professor relationship. Not all dissertation advisors function as mentors, nor are they necessarily suited for that role. The doctoral supervisor's role will change over time as the student makes the transition from student to independent scientist. Should problems or incompatibility develop, both people should have options to resolve them. If the protégé has a broader network of mentors, then these particular changes can be more easily navigated.

At this stage, protégés should pay special attention to their mentoring needs in developing a strong background in science, planning time for research exposure, and being sure not to adopt a myopic view of their career outlook.

5.3.2 At the Postdoctoral Level

Mentoring at the postdoctoral level can lead to productive working relationships that benefit both mentor and protégé, despite distinct career goals. When researching prospective postdoctoral supervisors, postdoctoral fellows should determine where previous junior members of the laboratory have gone. Their success can be predictive of positive outcomes. At this career stage, peer-to-peer mentoring is a strong possibility in the relatively new postdoctoral associations. Here, rich networking and mutual support opportunities are strong features. For the postdoctoral protégé, personal reflection is an important part of the process, and should include an evaluation of one's aspirations over the course of the training. It is extremely important to keep your mentor(s) informed if your career goals have changed, so that they can provide the type of guidance you may need.

Developing committed mentoring relationships is vital for all scientists throughout the progression of their careers. But few stages are as needful of guidance as the postdoctoral level, which is by definition a time of transition for most young scientists. While graduate students have a dissertation committee and other formalized support structures, postdoctoral fellows rarely have equivalent resources, even though their positions are considered to be training in nature. At many institutions, postdoctoral fellows fall between the designations of student and employee, and may be completely at the disposal of their principal investigator. This frustration is exacerbated by historically low funding levels that can put an academic career, where tenure is attained in part through successful federal grant applications, out of reach for many. Making matters even worse, many postdoctoral fellows find that their university training did not prepare them for, or even adequately educate them about, what other opportunities are available. So whether one is planning on continuing work at the bench or is remolding skills to suit a non-traditional career mode, the postdoctoral experience should ideally be crafted as a springboard to a permanent position.

Although research productivity is certainly necessary for success at the next career juncture, in no way is it sufficient to help one locate and land the dream position. Locating great mentors can be challenging because what it really entails is developing professional friendships with people who understand one's concerns and constraints, can advise on the state of the current job market, and ultimately have already overcome the obstacles that lie in the postdoctoral fellow's path. Finding the right fit in a mentor is a matter of chance, but postdoctoral trainees can stack the odds in their favor for building their own "professional advisory boards" if they are proactive about career development. The first step is to realize that multiple mentors are required to achieve different perspectives on one's goals. Informal mentoring relationships are usually found serendipitously, and many graduate students and postdoctoral fellows first look to their advisor and go no further. Unfortunately, often a principal investigator cannot be an adequate mentor for most postdoctoral fellows, given the unabated university focus. They may be out of touch with the needs of certain vocational paths, or postdoctoral fellows may not feel comfortable sharing their aspirations with someone who is largely an employer (Horvath 2006).

Finding a set of career-energizing mentors starts with being self aware, that is, realizing that you, the postdoctoral fellow, are in charge of your professional development, and if the proper support does not exist in your work environment, you must actively seek it. Participation in local professional organizations is one of the best ways to meet new individuals that can offer vital advice and direction. Local chapters of professional disciplinary based societies, as well as the Association for Women in Science (which in particular has been a leader in this area) develop their own mentoring programs and host both workshops and social events for members to network. Additionally, one's local postdoctoral association often holds seminars, career fairs, and various other events that can connect postdoctoral fellows with prospective mentors in the larger scientific community.

Although participation in professional society activities may detract from time at the bench, organizing the events and programs places one on the frontlines with prospective mentors. The shared goals inherent in event planning can act as ice breakers for those shy in approaching new friendships. Participation is also an excellent way to meet other postdoctoral fellows, who represent an underappreciated support group and source of mentors. Trainees just a few years beyond one's current position are often actively looking for a permanent job, and have plenty of advice for those fresh out of the Ph.D. gates. After all, one's colleagues today are also one's networking contacts tomorrow.

External, structured mentored programs can provide an important supplement to these professional friendships. Formal mentoring has the advantages of established goals, tailored design, open access, and "jumpstart" activities when progress in the relationship starts to reach a lull. Protégés should continually monitor their expectations and needs, so that they receive the advice and mentoring at the time they need it. Postdoctoral fellows need to view their resumes as a personal advertisement tailored to the job at hand, and not simply hope that they will be hired solely based on demonstrable technical expertise.

5.3.3 At all Post-training Career Levels

Having mentors who come from the ranks of institutional colleagues or professional society or agency committees is a logical career enhancement strategy. Now that one has completed preparation for the career, these individuals can serve an important mentoring role. Such aspects as sharing issues and situations with colleagues for advice and perspective constitute a form of mentoring. Using one's mentors to verify facts, ask about prevalent rumors in the field, or identify changing circumstances constitute very important reality checks. Because trust builds up over time, mentoring relationships developed at this career stage may last over many subsequent decades.

"Reverse mentoring" has often proven of great worth in those circumstances where more junior colleagues can provide critical perspectives, providing that the more experienced person is willing to listen for what is actually being said. In exhorting staff, students, or peers to push the boundaries of what is possible, one may glean important insights when one is "told what I need to hear, not what I want to hear". In these cases, the importance of mentoring upward is evident. That is, individuals in earlier stages of their careers can and should mentor those in later stages, provided that the relationship is an open and honest one. A clear sense of boundaries and sensitive topics is critical for both parties to understand. Such mentoring relationships can be supportive of scientists who wish to be honest about themselves and their careers.

Across the board, soft skills development is emphasized as one of the primary inadequacies inherent in the master-apprentice model of Ph.D. creation. That is, it is essential not only to be technically competent in one's chosen field, but also to learn and integrate the many effective ways of "working and playing well with others". One must attack career development and advancement with as much gusto as is usually given to chasing after scientific questions. This strategy includes having the self-confidence to find mentors who can provide critical career guidance when most needed. By being aware of one's resources, an array of both formal and informal mentoring options can nourish the protégé's career development. The astute professional will find that successful individuals from many scientific specialties are particularly eager to advise those in early phases of their careers. The key is to engage in activities that unveil new professional environments for networking, which almost always necessitate stepping away from the laboratory milieu.

5.3.4 Points to Ponder

- Reassess how you were mentored at your early career stages.
- Identify a mentor who can "follow you through time".
- Consider participating in a formal mentoring or career enhancement program.

Chapter 6
Career and Life Transitions

6.1 Work–Life Balance

> My inspiration to work with promoting women in science came from being a woman in a company and realizing that so few women were able to stay working and progress in their scientific careers.
>
> Joanne Kamens

Many women in science and engineering fields have at one time confronted the Superwoman Syndrome. The symptoms are well known: crippling anxiety over trying to juggle the roles of parent, spouse, community volunteer, domestic goddess, and high-achieving, career-driven dynamo. When protégés may be trying to wear all of these hats, mentors are an invaluable resource (and at times a lifesaver) on issues of work–life balance. They can offer perspective and suggest strategies honed from their own struggles. Unfortunately, mentors who can effectively speak about work–life balance are not always available to everyone. In these cases, early career scientists in need of guidance should consider or seek mentors outside their geographic or professional scientific area.

Work–life balance is of course a highly personal issue, and one person's solution may not be another person's solution. However, there are a number of commonalities in both defining the issues and developing time management approaches that are worthy of consideration for all scientists and engineers.

6.1.1 Finding Time for the Other Things in Life

Mentors overwhelmingly agree that the single greatest mistake early career scientists make is not arranging for enough personal and family time in their lives. Stepping away from the bench or computer becomes unnatural and can be accompanied by feelings of guilt or sloth, particularly for individuals who go through long training periods. One mentor discloses, "All of my protégés have had to deal with the guilt aspect of free time. Taking time off to travel or have fun with friends is stigmatized. I make sure to tell them that I didn't take time off when I was in graduate school and wished that I had later." Dr. Amy Lovell, associate professor of astronomy at Agnes Scott College, reaffirms this position. "I think the greatest mistake is exerting too much effort on work at the expense of other aspects of life, assuming that one will have more time later for spending with family and friends. Once overworking patterns are established, they are hard to break." Another hard truth is that refusing to take this time (or refusing to believe that making time is possible) can compromise one's long-term productivity. It is easy to find studies and news articles demonstrating that after about 8–10 hours of work, one's ability to work efficiently and accurately decreases drastically (Horvath 2007b).

6.1.2 Managing Your Employer's Expectations and Your Own

In the business world, the proper attitude is more important than ever. Although pride in one's work is a necessity for job satisfaction, achieving balance can be affected by one's attitude toward the position and misunderstanding of one's role in a corporate environment. Dr. Tom Wilson, a MentorNet mentor and a senior chemical research and development engineer, cautions, "When I started my career, I made the choice to totally devote myself to the company. I did this in the belief that the company would appreciate and reward me. But I soon discovered that 'business is business' and it does not matter how dedicated and successful you are, or what you've been told by superiors. If for business reasons you are no longer needed, then you are gone. Once you separate your worth as a person from your job, then striking a work–life balance comes more naturally. This lesson, which took me two times to learn, is what I try to pass on to my protégés." "I didn't learn my lesson either," admits another MentorNet mentor who wishes to remain anonymous. "I've played *Wonder Woman* several times in my career. The problem is that one can do it for a while, but eventually, there comes a time where one can't sustain the role. In the business world where the idea of continuous improvement has taken hold, last year's performance becomes this year's base standard. The fallacy of this is that continuous improvement is about incremental changes. But to become W*onder Woman*, 2–3 step changes must be made and sustained each year."

It is clear then that managing your boss's expectations can be just as critical as managing one's own if work–life balance is to be struck. Part of doing this is to aim to be a living example of a balanced work ethic and dispel the myth of "workaholism." One may begin by simply refusing to participate in the stress-inducing head games of petty one-upmanship such as, "Who left latest last night?" and, "Who took the shortest vacation?" A mentor at a national laboratory reminds us that "because of the perception that the number of hours one works counts more than accomplishments, one must have a 'tough skin' to stick to a shorter schedule and believe that ultimately the results will be what counts. One must have the self-confidence to ignore what others may think" (adapted from Horvath 2007b).

6.1.3 Strategies to Attain Balance

Although it is clear that taking a mandatory time-out is essential to one's well being, there still remains the question as to how to consistently ensure this will happen. One simple method is to set aside at least a half-day a week that is entirely work-free; another is to keep to a ritual of not taking work home. But this can be difficult for those of us who are unaccustomed to the idea or have erratic work schedules requiring evening events or meetings. A second option proposed by numerous MentorNet mentors is to at least build structured activities into the week, such as social groups, exercise, or classes, which require fixed appointments. One graduate student elaborates, "If I set up a personal training appointment at my gym for 7:00 PM on a week day, then I know I have a fixed amount of time to complete my work and I am not able to cancel last-minute without losing $30. This keeps me working very efficiently throughout the day but at the same time puts a reasonable cap on the total work volume" (Horvath 2007b).

Ann M. Viano, associate professor of physics at Rhodes College, advises, "I suggest that my protégés separate work and personal time by setting daily goals to be accomplished and focusing only on those goals. If they achieve them, then when it's time to go home for the day, they can wholly concentrate on family, friends, or other activities." It is true that these approaches cannot solve the problem of having more tasks and bigger goals than available time. But the simple fact is that the eschewal of excess labor can either occur in a planned manner or a stress-laden, last minute rejection. Most would agree that the former strategy certainly gives one a feeling of greater control over life. Viano adds, "There are solutions to almost every work–life balance problem, but you really need to find the courage to ask for help, not the time and energy to just complain about it all. Ask your colleagues for tips on preparing for class efficiently even though you're running on no sleep because you have a newborn at home. Ask your supervisor if your workplace has ever considered spouse job-sharing if you and your spouse are trying to solve the dual-career problem in the same field. Ask family to visit and help out at home during busy times at work" (adapted from Horvath 2007b).

6.1.4 Recognize That Balance Is Not Always Attainable

Most mentors confide that even their most well-rounded colleagues have times where balance is unattainable, and this has to be recognized and abided. The lack of balance during acute periods when dissertations are written, grants are prepared, and critical experiments are completed does not make one abnormal or a failure. But one may need to seriously reevaluate one's career path if non-work time becomes constantly unattainable. Dr. Lovell emphasizes, "I believe there are career tracks that are all work and expect the full devotion of the scientist at the expense of all other aspects of life. I have chosen a path where I can succeed, but at my own pace, and I encourage all early-career scientists to start at the very beginning with a model where they have time to do the things, both professionally and personally, that they value most. For some people, success is defined with respect to the all-work-no-play career, and this either leads them into those career paths or leaves them feeling torn if later on they desire changes in favor of balance."

No matter whether a scientist has found a workable balance or stands at a rough spot career-wise, a mentor can be a sounding board for frustrations, a provider of strategies, or a cheerleader when goals have been met. But it is always the protégé's responsibility to seek help and develop these relationships. If mentors are from different fields, then a protégé can feel free to convey "what is in my head and life without worrying that any of my words might reach the ears of a hiring committee or my bosses." Having a positive attitude regarding the feasibility of work–life balance is the most important initial step of achieving it, even if the answers are not immediately obvious or palatable at first examination. Reach out to your mentors to help you uncover these solutions.

6.1.5 Points to Ponder

- Seek out a mentor who has dealt with work or life issues similar to those that you are facing.
- Recognize that perfection is impossible to attain.
- Reassess your mentoring needs.

6.2 Coaching or Mentoring? What Do I Need Now?

> Success is to be measured not so much by the position that one has reached in life as the obstacles which he or she has overcome while trying to succeed.
>
> Booker T. Washington

A common trait of the most successful scientists is an astute management of their own strengths and weaknesses. For some of these lucky individuals, this self-awareness may have crystallized naturally. But many scientists, even individuals already well into their careers, require help in goal definition, prioritization, strength assessment, and plan development in order to attain the greatest efficacy in their professional lives. At one time or another all scientists have found mentors along their career paths. However, there is another personal development paradigm that can be highly effective, although it is deployed in an entirely different fashion. Professional coaching is a service where a coach and a client collaborate to develop a highly individualized program for increasing job performance and reaching career goals.

A quick investigation into the distinctions between mentoring and coaching will reveal that this is a highly debated topic. Some individuals feel as if the two terms are to be used synonymously, whereas others see them as completely separate paradigms. But a closer examination reveals numerous important distinctions. Lois Zachary, a consultant with Leadership Development Services, gives a mentor's perspective on the difference between mentoring and coaching: "Coaching is always part of mentoring, but coaching does not always involve mentoring. Coaching within the context of a mentoring relationship has to do with the skill of helping an individual fill a particular knowledge gap by learning how to do things more effectively" (Horvath 2007a).

"While mentors may use coaching techniques and coaches may provide mentoring in some situations, coaching and mentoring are not the same thing," asserts seasoned management coach and consultant, Moira F. Breen. Breen earned a master's degree in biochemical pharmacology and embarked on a 15-year career in research and development in the pharmaceutical industry, which included international positions with GlaxoSmithKline. Knowing all too well the demands that scientists face in both their professional and personal lives, Moira created a coaching consultancy to aid individuals with career transition and management and leadership development. While she views mentoring as either a volunteer activity or part of a strong interpersonal relationship between two individuals at different career crossroads, coaching is uniquely an occupation. "Coaches receive training in techniques and processes designed to support the development of others, and they have chosen this work" (adapted from Horvath 2007a).

Although coaching is still a relatively young profession, standards have been developed and are being adopted quite rapidly to protect clients and ensure the quality of coaching clients receive. Bodies such as the International Coach Federation guidelines, devise codes of ethics, and develop certifications and other credentialing metrics for professional coaches. While coaching comes with the aspects of accreditation, mentoring is unique in that the mentor and protégé often share vocations and therefore can advantageously discuss very specific, technical issues pertinent to their common field. Often there is a master and apprentice relationship, and this phenomenon does not have a regular counterpart in coaching.

A coach is not 'senior' to the client, and in fact, it is not even necessary that both individuals come from the same field. Moira Breen states, "It is not necessary for the coach to have carried out the specific role and responsibilities that the client performs or wishes to develop because coaches are not training their clients. My clients have included lawyers, dentists, veterinarians, architects, and directors of non-profit organizations, not only those in senior management roles in the pharmaceutical industry similar to my own experience." That being said, Breen does concur that many clients choose coaches with ties to their own field since they may be familiar with the work conditions and can provide expertise based on their own experiences (Horvath 2007a).

This is the point where coaching may overlap with mentoring and consequently create confusion in distinguishing these two tools. Breen clarifies, "A coach may appropriately provide these services to a client but they are not strictly 'coaching,' and the coach should be aware that they are stepping out of that role to meet a different need." Breen explains that coaches do not use their own journeys as a starting point for client development, but instead work with the client to elicit creative strategies to meet their goals (Horvath 2007a). Mentoring, in comparison, draws more heavily on a mentor's experiences given that the career path being discussed is usually shared by both participants. This dissection of these two career development models underscores their complementary natures, for each fulfills different needs.

6.2.1 Embracing New Communication Paradigms and Developing Priorities

This need to embrace new communication paradigms, however, is one area where mentoring and coaching find firm common ground. Although mentoring is traditionally performed in person and associated with some element of on-the-job training, new models have arisen that embrace all that internet technology has to offer, as noted in Section 5.2. Despite the fact that the information age has created a multitude of gadgets to remind scientists to nurse their workloads, priority management is a challenge regardless of the decade. Mentors can uniquely provide protégés with knowledge and domain-specific suggestions, but a coach attacks the problem in a different yet complementary manner. Moira Breen discloses, "I help clients on moving towards completing the things that are important to them instead of the things that are not. I help clients develop strategies to handle anything they decide to handle but nothing that is highly specific to keeping up with advances of knowledge in their field." In other words, coaches concentrate more on eliciting what is really important to the client and assigning priorities in that manner.

Coaches can help with procrastination by getting to the root of a client's problem – overcoming fear. Experienced coach Jane Chin has noted that "People may not always be attuned to how fear drives their indecision, even though they may hear about fear all the time. When a client comes to me and says, 'I have a problem with procrastination,' I find after discussing further, there is fear hiding behind the procrastination. . . .The problem is that either we may not always be aware of the 'right' prioritization best-aligned with our values, or, we know, but we are held back by fear." She emphasizes that prioritization can be a very difficult task and is not merely a matter of acting tactically to define a 'to-do' list bracketed by time and resources. Scientists must think strategically in prioritizing in order to meet career goals. "This is where I've found coaches to be very helpful. I've personally worked with coaches as sounding boards for strategic input and for holding me accountable to the tactical output." (Horvath 2007a).

Although coaching and mentoring have many areas of overlap, their modes of execution are quite distinct. Coaching services seek to elicit one's priorities and goals through a series of self-discovery exercises, and the coach should not *a priori* have a set opinion on what those goals and priorities should be. Mentoring, on the other hand, is a process whereby the advice and direction given to the protégé should take advantage of the mentor's own experiences. Another key difference between the two is that while a mentor and protégé pair often work within the same general field, coaches and clients can sometimes operate most effectively without this commonality. Such an arrangement forces the client to leave the comfort zone of shared technical ground, which as a result may elicit career-shaping strategies and tactics that may not have been apparent if both individuals utilized the same domain of knowledge. But ultimately, it is up to the scientist to decide whether one's path could benefit from professional coaching as a complement to the myriad mentoring relationships that are built over a career.

6.2.2 Points to Ponder

- Assess whether your current strategies are meeting your career development needs.
- Explore career development and enhancement tools that complement your mentor's perspectives.
- Question the "conventional" wisdom.

6.3 Transitioning into a Different Career Pathway

Nothing in life is to be feared. It is only to be understood.

Marie Curie

As one gains experience both as a professional and in the larger context of life, the moment may arise when taking a different career direction appears the best option. The experience of Jane Chin is instructive in providing food for thought that protégés can ponder as they look introspectively to identify critical issues in their own quest. She has stated that she chose her early career path by "never thinking about what I wanted to do or what I enjoyed doing. I had taken the road of least resistance. I've learned the hard way that achieving full potential requires both intimate self-knowledge and being in the right 'space' or field to exercise your talents. If I had asked myself a few key questions early in my career, I could have spared many years of trial and error on the way to achieving my full potential" (Chin, 2008).

Table 6.1

Key questions for a change in career pathway

What are my strengths?
What do I really want out of a career path?
How do I make my knowledge and experience work for me if I embark on a different career path?
How do I gather transitional momentum?
How do I create opportunities?

Jane has identified some key questions that protégés should explore. Although she articulated these in the context of an early career stage, they are equally applicable for anyone at any time in her professional career. She articulates important concepts that can be explored with a mentor. Below are expansions on her thoughtful analysis, mapped more closely to tools and techniques that individuals may deploy in seeking out and working with their mentors.

6.3.1 Know Your Strengths, Your Thought Processes, and Your Values

Your strengths are your natural assets. Your thought processes are the mechanics through which you may apply your assets to generate new fields of possibilities. Your values determine what gives you a feeling of meaning and satisfaction in a career. You may identify your strengths by taking assessments designed for this purpose, committing to periodic introspection, and working with a mentor.

6.3.2 Let Go of What You "Should" Want

If you have ever felt trapped in a career, then you probably were following a series of "shoulds" during your career. You may find yourself surrounded by negative reinforcements, where your peers or advisors continually remind you of these "shoulds." In this situation, you want to be very selective of the company you keep. Your support network should include a mentor or mentors who are supportive of your transitional aspirations. You may want to pair up with a peer who has similar transitional goals and support one another with peer-to-peer mentoring.

6.3.3 Make Your Scientific Background Work for You

One of the most common challenges that scientists face is the transition from research intensive scientific careers into non-research scientific careers. It becomes scientists' imperative to translate the diverse skills they have acquired in the process of conducting research and working in a laboratory environment into new situations where they may successfully apply their skill sets. When communicating with prospective employers, scientists need to learn the jargon or language that hiring managers recognize. In these circumstances, having a mentor more experienced in these arenas will guide you into a more thoughtful approach and analysis of the next steps to meet your goal.

6.3.4 Put a Price Tag on Procrastination

If procrastination is really a matter of time management, personal organization, or prioritization, then the wealth of publicly available seminars and scheduling tools should solve this problem.

However the source of procrastination is rarely a lack of knowledge in organizing time or priorities. The source of procrastination is fear. Fear of unknown, fear of failure, fear of public opinion – these are a few fears that plague professionals in transition and sometimes cut off their initial momentum. In most circumstances, these options are valid questions. Putting a price tag on procrastination is a constructive reality, and mentors can help hold one accountable for tasks that move you toward your ultimate goal.

6.3.5 Create Your Opportunities

Do you believe that you "have what it takes, but just need an opportunity?" If so, you join ranks with many scientists who desire a career transition, but assume that once their resumes are polished, their next step is to wait for their "lucky break." Ph.D. scientists are not unique in making this dangerous assumption; people in career transition can keep waiting to bump into the right people at the right time that will give them the right opportunities. Networking is an essential skill for creating opportunities. Reaching out to your mentor(s) should always be a critical part of your networking activities because networking is a skill of "connecting with the connectors". The better you become at networking, the more likely you're able to connect with people who may point you in the right direction and refer you to others who may have the opportunities you are looking for.

6.3.6 Points to Ponder

- Do not be afraid to choose new pathways when you are frustrated with your current career or life status.
- Reflect on what is most important to you.
- Take action steps to move yourself toward your ideal career path and lifework.

Chapter 7
Navigating Interpersonal Contexts

> *Every professional benefits from developing the skills to work productively with a wide variety of people, and this includes recognizing and taking appropriate action when behavior patterns disrupt the work environment.*
>
> Sara (Sally) Tobin

7.1 Inappropriate Relationships with Mentors or Supervisors

There are many reasons why it is not a good idea to have an 'amorous' relationship with a mentor or a boss. Consider the following:

The relationship is over...but the discomfort could go on if:

your mentor is the only professor who teaches a course you must take, or
you must attend the same departmental functions, or
you must still work on a committee together.

Although some people have managed to have a sexual or romantic relationship with a professor or supervisor without subsequent difficulties, that is the exception rather than the rule. The graduate student who married her professor is often mentioned as a model of some sort; however there are many more students who have been in relationships with professors that did not work out and subsequently the students changed their major, transferred to another school, or dropped out.

There are problems inherent in dating your professor (if you are a student) or a more senior colleague (if you are a professor or other STEM professional). Some of these are:

- When the person has power over your grades or your promotion or tenure, and hence your future, it is difficult to have a relationship of equals.
- Although most faculty members do not have romantic or sexual relationships with their students or postdoctoral fellows, those who do are typically involved in "serial relationships" (i.e. one relationship after another). Many of these individuals prefer younger partners to those closer to their age because those younger are less demanding and more dependent.

- It is difficult to keep such a sexual or romantic relationship with a professor (or senior colleague) secret for very long. When others know about the relationship, your good grades (if you are a student) or professional opportunities (if you are on a career track) are likely to be seen as preferential treatment. If you are a faculty member, the good work that you do, such as a published paper, is likely to be questioned and attributed to the person in power rather than to you. Thus the validity of your achievements in not taken seriously by peers and other colleagues in the department and often elsewhere as well.
- Any extra "help", good grades, or good evaluations from your mentor may make you unsure of the value of your own achievements. You may have self-doubt as you question whether your grades or evaluations would have been as good as they are if you were not involved in the relationship. Note that it is hard for someone in a personal relationship to evaluate the other person objectively.
- Other faculty, students, postdoctoral fellows, and colleagues may feel very awkward about the relationship and avoid you.
- If the relationship ends badly (and it often does, with hard feelings on both sides), depending on the power of the mentor and his/her position, the following may occur:
  - You might find it difficult to use the mentor as a reference.
  - The mentor might say negative things about you to colleagues and students.
  - Both people are likely to feel very awkward during any subsequent contacts, many of which cannot be avoided if they are in the same department or STEM field.
  - Often when a relationship ends, a woman may avoid her ex-mentor and his/her colleagues in the department and thus feel alienated. In fact, often the person without power leaves the department or organization because the discomfort experienced is so negative and extensive.
- Some previously consensual relationships turn into sexual harassment when one of the persons involved wants to terminate the relationship and the other person persists and uses threats or bribes in an attempt to maintain a connection.
- In many institutions, romantic relationships (especially between faculty member and student or between supervisor and employee) are frowned upon and in some cases prohibited. Conflict of interest issues, along with ethical issues and issues of trust and inappropriateness are typically cited as reasons why such relationships are not sanctioned.

7.2 Identifying Problematic Behaviors

It is reasonable to presume (and to expect) that all behavior of mentors should be above reproach. Nevertheless, there are circumstances when the protégé may have a vague sense of discomfort about the "tone" of a mentoring relationship.

To that end, Bernice Sandler (http://www.bernicesandler.com/) prepared the following checklist of questions intended to explore this subject.

She notes that some men (and women) have difficulty understanding which behaviors are considered as sexual harassment. The following questions can be helpful in assessing one's own behavior (particularly apt for mentors) or someone else's behavior (for protégés to assess possible concerns about their mentors' behavior).

1. Would I mind if someone treated my spouse, partner, girlfriend, mother, sister, or daughter this way?
2. Would I mind if this person told my spouse, partner, girlfriend, mother, sister, or daughter what I was saying and doing?
3. Would I do this if my spouse, partner, girlfriend, mother, sister, or daughter was in the room?
4. Would I be comfortable saying the same thing or acting the same way to my mother, sister, or daughter?
5. Would I do this if the parent, spouse, significant other, boyfriend/girlfriend of the other person was present?
6. When a person objects to my behavior, do I apologize and stop, or do I get angry instead?
7. Would I act this way if I didn't have power over this person, such as being a supervisor or executive? Do I have other kinds of power over this person, such as being bigger or with more status?
8. Is my behavior reciprocated? Are there specific indications of pleasure – not "she didn't object" but specific behaviors indicating that she is pleased by my behavior?
9. Would I mind if a reporter wanted to write about what I was doing?

It is notable to keep in mind that "if you have to ask", such behavior is likely to be high risk and it is probably better to not do it.

7.3 Defining Issues

Sometimes the sexual issue in a mentoring relationship, especially between a male and female, can be problematic. Both women and men, whether students or faculty may sometimes misinterpret each other's intentions or they may worry about their own behavior being misinterpreted. When such problems of this kind arise, women faculty being mentored by other faculty members or administrators are vulnerable because their careers may be literally at risk. Similarly, women students and postdoctoral fellows may find their future careers jeopardized. Students and faculty members alike need ways to make their own professional concerns clear, and institutions should have policies that clarify appropriate relationships. While no advice will work in all situations, the following may be helpful to many mentors and those whom they mentor.

For protégés and mentors:

- Meet in places that discourage sexual intimacy, such as departmental offices, laboratories, and other work-related settings.
- Always talk in a professional manner, whether you are discussing personal or professional concerns.
- Avoid sexual joking or innuendos, comments about personal appearance, and intimate confidences.
- If you are having lunch, dinner, or drinks together, invite one or more persons to join the two of you, especially in the initial phases of the mentoring relationship.
- Mention that you have a spouse, a partner, or significant other early in the mentoring relationship. This often has the effect of giving the message that you are neither available nor interested in a sexual or romantic relationship.

For protégés:

- Try to get to know your mentor's spouse/significant other and family, and if possible, talk about and introduce your mentor to your own spouse, significant other, or partner.
- If your mentor suggests a sexual or romantic relationship, confront the issue in a straightforward way. For example, say something like, "I'm very flattered by your attention to me, but I don't want to ruin the working relationship we have developed." In this kind of circumstance, it is wise to keep a record for your own use by writing a description of the incident(s). If possible, talk to other people who know the mentor well to see if the attempted sexual relationship is part of a pattern.
- If unable to discuss harassment problems directly with the mentor, you may find that a carefully constructed letter to the mentor may be helpful.
- If the mentor tells sexual jokes or stories or makes intimidating remarks with sexual overtones, you can respond in a number of ways such as:
    - stating in a straightforward manner that you find such remarks offensive.
    - pretending that you don't understand the remark or joke and ask the mentor to repeat it, and then ask the mentor to explain it.
    - pretending shocked outrage (the "Miss Manners" approach) by saying something like "I beg your pardon!" A variation would be "I can't believe you actually said that!"
    - using humor as in "Oh, is this a test to see how I would handle sexually harassing behavior?"
    - composing a letter, as mentioned above, would also be appropriate for this kind of behavior.
- If the behavior continues, you should seriously consider getting out of the relationship. Also, consider talking to the person who handles your institution's sexual harassment complaints to discuss your informal and formal options. Be sure to read the sexual harassment policy carefully.

- Ignoring sexual harassment usually does not stop the behavior. It often increases it because the harasser typically assumes that since there is no protest, the other person likes the behavior or is too weak to stop it. If sexually harassing behavior does not stop when the other person indicates displeasure, the behavior is likely to escalate.

For mentors:

- Call your protégé by name rather than by terms of endearment such as "dear" or "honey".
- You can leave the door open when you meet with others, regardless of their gender. Some people only leave the door open when the person they are meeting with is of the other sex. This behavior is discriminatory because it treats men and women differently. It also ignores the possibility of same sex harassment allegations.
- Should a protégé "come on" to the mentor, or otherwise indicate a sexual or romantic interest, make a very clear statement that you enjoy working with the student or junior colleague, and that you do not want to jeopardize the relationship or to violate conflict-of-interest rules. If the person persists, make an even stronger statement about the inappropriateness of this kind of behavior and that if it continues, you can no longer be part of a mentoring relationship. In this kind of circumstance, it is wise to write a description of the incident in a memorandum for your own file, and in some instances, to share the memo quietly with your administrator.

For institutions:

- Institutions should have conflict-of-interest policies that clarify appropriate relationships between faculty and students in their classes (or in the case of business and other institutions, between supervisors/managers and subordinates). Some institutions have policies that require anyone with professional responsibility for a student or employee with whom he/she also has an "amorous" relationship to report that relationship to a supervisor so that alternate arrangements will be made for someone else to carry out the professional responsibility.
- Institutions should have guidelines for mentoring relations which make it clear that sexual or romantic relationships are inappropriate.
- Institutions should assure that their sexual harassment policies allow for informal resolution of conflict-of-interest and sexual harassment complaints so that those harassed are more likely to report the situation.
- The sexual harassment policy should also have a statement prohibiting retaliation against anyone who reports sexually harassing behavior or who participates in a complaint process (for example, witnesses). Examples of what constitutes retaliatory behavior should be included. As well, any official who discusses a sexual harassment report with someone who is accused (formally or informally) of sexual harassment should mention the seriousness of retaliatory behavior.
- The institution's policy should state that a consensual relationship is no defense to a subsequent charge of sexual harassment.

7.4 Intervention Strategies

Many people find it difficult to intervene when they observe someone sexually harassing another person. Sometimes they do not know what to say or how to intervene. Sometimes they are worried about their peer's reactions, that they will not be liked or supported by them. Some people feel that "It's not my business". Others may not realize that in some instances they have a responsibility to intervene.

Ignoring sexual harassment, or other offensive behaviors, does not work. Not responding to the behavior gives the impression that the person observing it or the person to whom it happens is either too weak to deal with the behavior and/or both of these parties find the behavior acceptable, and/or the person to whom it happens likes the behavior and is not offended by it. Rather than being passive bystanders to sexual harassment and other inappropriate behaviors, faculty members and administrators and managers of scientific organizations need to develop ways to respond when they observe such instances.

If you observe sexual harassment of someone for whom you are professionally responsible, such as an employee you supervise or as a faculty member observing student-to-student harassment in class or related activities, you should always respond in some manner either immediately or shortly after the harassing behavior occurs. However, even in cases in which you are not professionally responsible for others' behavior, you may want re-respond. To some degree, *everyone* in a community is responsible for seeing that the community's standards of behavior are followed. Just as faculty members would in almost all instances intervene when they observed cheating, and would report plagiarism or theft of departmental materials, they need to respond when sexual harassment occurs.

What follows is a list of ways in which persons can intervene when they observe sexually harassing behaviors. The strategies will also work with other offensive behaviors even if they are not sexual harassment or are not prohibited.Not everyone is comfortable with all the strategies described. The aim is to provide a variety of ways to respond so that individuals can choose the strategies with which they are the most comfortable. Many of the following techniques are also appropriate for individuals to use when they are being harassed by a coworker, supervisor, or student.

To the extent possible, interventions should occur immediately after offensive behavior occurs.

Humor: Using humor and playfulness are good ways to handle harassment (if you can think of something immediately) because humor in a stressful situation connotes strength. Humorous comments demonstrate that the person making the clever comment was not overpowered by the harassing behavior. Humorous comments also help to break the cycle of behavior. Unfortunately, many of us think of wonderfully funny comments later, when it is often too late. Here are some standard remarks which, when said lightly and jokingly, might be helpful in a variety of situations. They could also be said in a direct manner without humor.

- I would hate to hear that you are being sued by students for sexual harassment.
- So, is this a test to see how I would handle sexual harassment?
- Are you sexually harassing me (or name of person or group) again? I am going to have to call the sexual harassment office right now!

Surprise: Showing surprise indicates that the behavior is not acceptable.

- I beg your pardon!
- I can't believe you said that!
- I'm absolutely speechless!
- Do you know that could be seen as sexual harassment?

Direct Response: It helps to name or describe the behavior that is inappropriate.

- That comment is offensive to all of us; it is unprofessional and probably is sexual harassment. That behavior has to stop.
- This is not the first time you've said things like this which many people would call sexual harassment. It's getting in the way of your effectiveness.
- That behavior is not acceptable here. It violates our policy on sexual harassment.
- That behavior is disgusting (unprofessional, immature, inappropriate, etc.)

Pretending to Not Understand: In addition to the suggestions and options above, one can pretend not to understand a particular joke or remark, particularly in response to sexist remarks, jokes, and stories which portray women and racial or other groups as the object of laughter or ridicule. Most of these jokes are offensive to these groups, though individuals may not openly complain. Keep a deadpan expression and state that you do not understand the remark and ask the person to repeat it again, using one of the statements below. Follow up by asking the person to repeat again whatever they just said, and continue to claim that you don't understand what they mean.

- I don't get the point of your remark.
- I don't understand what that means.
- I don't understand how your comment is relevant to our discussion.

Private Reprimand: If you do not want to address the behavior publicly, you can still intervene immediately by addressing the person who behaved offensively, stating that you would like to talk to him or her privately. This should be done in a very serious tone of voice so that other bystanders receive the message that the person's behavior was out of line and that it will be dealt with. This works best if you have higher status or power (such as being a supervisor) over the person who in engaging in inappropriate behavior.

Sending a Letter and a Copy of the Policy and Other Materials About Sexual Harassment to the Person: Sometimes it is helpful to underline or highlight that part of the policy which has been violated. If you do not feel comfortable sending the policy under your name, ask the person in charge of the sexual harassment policy to do so.

Writing a Letter to the Perpetrator: This technique has been an extraordinarily successful method for dealing with sexual harassment as well as other forms of interpersonal conflict and inappropriate behavior. Such a letter consists of three parts:

> Part I. The writer describes what happened in a very factual manner without any evaluative words. Usually people agree about the facts but disagree about the interpretation of those facts. What the letter does is separate the facts from the feelings. Statements such as the following are examples:

Last week at our staff meeting you made several statements about the inappropriateness of women becoming engineers.

Yesterday at lunch you stated that Professor Mary Smith was acting 'just like a woman' when she criticized your suggestion. On several occasions you have made similar comments about other women faculty and students.

> Part II. The writer describes how he or she feels about the incident(s), again in as many words as necessary, and in non-evaluative terms.

I am very upset with this behavior. I find it offensive and disgusting.

Your behavior makes me feel very upset.

I find this behavior unproductive and it interferes with the productivity of our meetings.

I worry about the impact these remarks will have on morale.

I am concerned that someone will file a formal complaint against you (or our unit).

> Part III. This part is usually very short, noting what the writer wants to have happen next.

I want this behavior to stop at once.

I want you to treat your colleagues (students, assistants) in a professional manner, the way every colleague (student, employee) has a right to be treated.

Typically, the letter is sent by certified mail, return receipt requested, which is a way of informing the recipient that this is an important letter. Should the harassment continue, the letter can be used as evidence that the perpetrator was informed that the behavior was unwelcome. The person keeps a copy for him or herself, but does not send a copy to anyone else. The letter works best if it is a private communication between two individuals. If the letter is copied to others, the recipient of the letter might go to those copied in an attempt to destroy the credibility of the writer.

Such a letter is successful 90–95 percent of the time. It will not work with a very hostile person or someone who is sadistic or intransigent or with groups of harassers. Most of the time the harasser says nothing after receiving the letter, but stops the behavior. Once in a while the harasser wants to apologize or explain, but it is best not to get into a discussion of the behavior but to simply say, "I am not going to discuss it. I just want the behavior to stop."

Report the Behavior to the Appropriate Person: This strategy is particularly important if the person who is engaging in harassment is unresponsive to your efforts and the behavior persists, or if you are uncomfortable dealing directly with the harasser. Reporting the behavior in person to the individual in charge of the institutional policy is most effective. You can also bring others with you who may have observed the behavior, have knowledge of it, or are troubled by it. The person in charge should be able to tell you what steps they plan on taking, such as a formal letter from the institution or a formal talk with the person. If they cannot, you may want to consider going above that person to someone with greater authority. You may also want to write a follow-up letter to the person to whom such behavior is reported, asking to be kept informed about what steps have been taken. You should also inform them about whether the behavior has continued or stopped.

Keep records about any incidents you observed, what you or anyone else did, what the response was, etc. A private memo to your own files is a useful documentation, should it ever be needed.

Chapter 8
Starting Out with the Right Education

8.1 Compelling Voices

> Tell young women that they can do it, that they are no different now than people who made important contributions were in their own youth.
>
> Vera C. Rubin

As part of an initiative funded by the National Science Foundation, the Association for Women in Science compiled an extensive database of in-depth interviews with 26 women scientists, engineers, mathematicians, and administrators and 16 postdoctoral fellows and students. While this group clearly is not a complete and broadly representative group of practitioners and scholars, it does present a cross section of these women's views of the joys and sorrows, problems and solutions, and strengths and weaknesses of the STEM professions. This chapter and Chapter 9 address the themes that occur and reoccur.

Before turning to these themes, however, a few statistics about the respondents seem in order. Among the students are two postdoctoral candidates, eight graduate students, and six undergraduates. Student respondents, mostly from the western United States, come from five racial/ethnic groups. But how to capture this diversity? Ethnic and racial designations, always a sensitive matter, are also complex. Is a woman black, an African-American, or of Afro-American extraction? Is another Hispanic, Latina, or Chicana? Arguments can and are advanced for – and against – all such descriptions. Like women in the STEM professions as a whole, the students cluster in the life sciences, although there are a number of women studying engineering and the physical sciences.

The scientists are an even more diverse group professionally, ethnically, and in terms of seniority. Although respondents often defy single classifications, partially because of the rich complexity of science as practice and profession, the women interviewed fall into these broad groups: About six devote themselves primarily to research; six, to administration – from a grants officer to a university president; seven are academic scientists; eleven are mathematicians or engineers; and two mental health professionals. The categories, however, are misleading; they inevitably require generalization and simplification.

Classification according to field is similarly slippery; however, broadly, the 42 respondents' focuses are as follows: 18 in the life sciences (half in medical fields, half in biology or combining fields such as biotechnology); seven in chemistry; eleven in mathematics, engineering, and/or computer science; and six in physics/astronomy, environmental science, and science education.

As each woman explains her perceptions about what factors encouraged her and what discouraged her – some themes reoccur, but much is uniquely individual. The impetus to move into fields often empty and intolerant of women and the courage to remain in them has three touchstones: Madame Curie, mentors, and stubbornness. A handful of respondents see in the life and work of Marie Curie a model and a confirmation of the possibilities open to them as women and as scientists. While the great French scientist stands in some ways as a distant mentor as well as a model, many women who chose science mention specific guidance from family members and teachers – the mentors in the flesh that historic figures and paper guides like this handbook can help to support.

Finally, those interviewed cited their own stubbornness – or "determination" or "inner-directedness" or persistence – as essential factors in their professional choice and continuation. To quote one of the scientists, those who wish to enter science should "stay the course, putting up with whatever is necessary to get through the degree first and then into the first job." In a slightly different vein, a research associate in astronomy explains, "I know I'm not brilliant. My mind is slow and steady, but I know that if I bang my head against a problem, I'm going to solve it. Just let your brain integrate the information and finally come up, with the solution."

8.2 Informal Education

> We women in science careers want to see girls become aware that education can be a key to a better life through planned careers dependent upon choice rather than chance.
> Betty P. Preece

Most of those interviewed view the mentoring relationship as one often vital link in the whole process of education, which occurs – or can occur – at any time and in any place. Although formal education is important and was discussed by many respondents, most learning takes place outside classroom walls. A child's first teachers are usually her parents, and many respondents cited the support of family members as being crucial in leading them toward science and mathematics. Others spoke of teachers and other mentors; a few, of networks.

A number of respondents, however, found their early environments strewn with roadblocks to science. Parents and relatives often believed that scientific activities, much less an eventual scientific career, were not appropriate pastimes for girls. One scientist reports that the opposition from her mother "who could not understand that I did not want to be the wife of a vice president. I wanted to be the vice president" caused lifelong damage to their relationship. Another, who remembered being preoccupied with mechanics as a child, mused that "sometimes my mother was irritated by my interest in machines, rather than dolls." Almost without exception, future scientists growing up in home environments hostile or indifferent to science had their interest awakened by gifted teachers.

In contrast, many girls discovered their passion for science through the enthusiasm and guidance of parents and other relatives – grandparents, aunts, siblings, and the like. One respondent came from a "family of scientists," and she suspects that she and her sister, also a scientist, "may be the only cooks who, when learning to make a cheese soufflé, were instructed to add the hot cheese slowly to the egg yolks so as not to denature the protein." A future ecologist remembers reading science books with her father as a child, "when I was sick, we would read a book about what was ailing me at the time."

Early interest in nature, on occasion sharpened by growing up on a farm, often led to a permanent commitment to science. A biologist remembers growing snails in her room, much to the disgust of her mother. Another young scientist was inspired by her experiences with school science fair projects. She recalls her award winning project, "From Eggs to Chicks" where her eggs hatched in the middle of the fair.

8.3 Precollege Education

Never give up, because the hardest things in life sometimes bring the greatest rewards.
Jacqueline Zaldana

In a book on achieving effective mentoring, it is fitting that so many respondents in the Association for Women in Science study (particularly those coming from science shy backgrounds) cite their schoolteachers as among the important reasons they went into science. Nearly all mention a teacher at some level – from kindergarten to graduate school – with the large majority remembering positive experiences. And, among the women who received discouraging treatment from some of their science instructors, two of them simultaneously praise the work of others. High school teachers receive the most acclamation. One woman, who became one of America's pre-eminent chemists, attributes the birth of her career entirely to the work of a high school chemistry teacher. At the same time, she recalls with distaste certain professors who did not want a young woman, no matter how talented, in their laboratories.

Much of the current recent reform in science education has focused on the elementary school and middle school years, and a number of the women interviewed discuss the vital need that it continue – for example, noting that minority and female exemplars come into classrooms and that children learn more mathematics earlier. Nonetheless, the appreciation that nearly half the respondents show to their dedicated and talented secondary teachers indicates the vital importance of high school education in encouraging fledgling scientists.

In spite of their gratitude, only one of the respondents had worked as a precollege teacher, although about a third had in the past and/or were currently teaching in institutions of higher education. In fact, a number of the scientists resented being pushed, by family or other teachers, toward the high school or elementary school teaching, which, in their youth, was an "acceptable" profession for a woman. The reasons for this gap are various; one certainly being chance, but perhaps the lack of teachers among the respondents is in some ways indicative of an American anomaly. Although we as a nation pay lip service to the importance of precollege teaching, particularly to science and mathematics teaching, the profession lacks prestige and, of course, funds. One science educator cautions "When scientists who have 'made it' in their scientific disciplines decide to make pronouncements about science education, the science community listens to them. If scholars get Ph.D.s in science and go straight into education, they should have their eyes open. There's not as much prestige in education as there is in being a research scientist."

Although she is speaking of university-level science educators and not precollege teachers, her generalization, unfortunately, applies to both.

Several women had negative experiences during their precollege years. An African-American engineer recalls being discouraged by her math and science teachers. She says,

> I remember I took a math enrichment course one year and I was very excited about a formula we learned. I brought it to my math teacher at school, and he told me to 'get a life'. Or I would ask my teachers for extra problems to do in class, and they would call my parents and tell them I thought I was better than everyone else.

Several student respondents commented on the failure of their high school to counsel them on postsecondary educational options. High school counselors did not provide adequate information about colleges; one high school counselor actually gave clearly inaccurate information, telling her students that MIT did not accept women!

8.4 Undergraduate Education

> Never be intimidated by statistics. If you have determination and passion, you can go far.
> Sheina Yellowhair

The role of teaching, and of role models, at the undergraduate level is a critical aspect of stimulating young women's continued interest in science, but it can present a dilemma for the female faculty member. One interviewee stated "Women are usually good teachers. Department chairmen say 'You are so wonderful with the students, I want you to continue.' The trick is not to be good at it: The person who wins the teacher-of-the-year award never gets tenure at the more prestigious schools." Whether this scientist is literally right or not, her opinion is not unusual.

Nonetheless, while few respondents called their relationships with their mentors and teachers at the postsecondary level formative in helping them choose science – by their undergraduate years, most had already found their vocation – many report warm and important relationships with professors. One woman, for example, after having dropped out of a science major so that she could take more liberal arts courses, had her first disheartening experience in a chemistry class of 1,000 – probably an example of the "weed-out" introductory science courses that many have criticized – corrected by a galvanizing woman bacteriology teacher. The former dropout is now a preeminent marine biologist.

A small percentage of the scientists attended women's colleges; three of those four found their experiences supportive and important in their decision for science. The fourth was less positive, noting both "pros and cons." She enjoyed the small classes her women's college offered and one-on-one contact with many of her teachers. And she liked seeing women faculty and students in leadership positions. But the experience, she said ruefully, had its drawbacks. Her college "modeled an ideal, not a real world," she explained. When she entered the professional world, it was a shock to find so few women in mathematics and engineering. She had moved to an environment where men were mostly in charge – just the opposite of that to which she was accustomed. In this regard, it is notable that much consideration is now being given to ways in which coeducational colleges and universities could better encourage women in science.

Three of the six undergraduates interviewed mentioned the challenge of surviving the difficult introductory courses, particularly mathematics and chemistry, required for the biology major. They all advised high school girls interested in science to persist, even if the work at first seems difficult. Presumably, they will follow up on their own suggestion. A government mathematician notes that once factors such as gender, ethnic background, and level of education are discounted, she said, salary differences are explained by the number of undergraduate mathematics courses taken. One woman summarized, "You give up a lot when you're in science because of the time you have to put into school, but my sense of accomplishment seems worth my sacrifice."

Several of the undergraduates interviewed attended the University of California, Irvine (UCI) and are members of minority groups. The students extolled the value of the support group for undergraduates in biomedical fields at their institution, the California Alliance for Minority Participation in Science (CAMP). This program focuses on preparing minorities for the transition from high school to college. CAMP also provides a network of friends and study partners throughout the school year. One student expressed the benefit of the CAMP program, saying "...there are lots of people who don't look like you. I was the only Latina girl in my physics class in high school, so I felt alone all the time. The students I met through the CAMP program have become a support system for me in college; we sit together in classes and tutor each other."

Many women recently out of college stressed the importance of doing internships and/or getting research experience while an undergraduate. The internship experience was especially important to the engineering students, who felt that they wanted to get a feel for the career before they dedicated themselves to it. All of the current undergraduate students expressed an interest in getting research experience while they were still undergraduates, saying that "it gives you a perspective on the field and helps you develop your interests."

8.5 Graduate Education

Believe in yourself – be your own biggest fan.

Anne Bentley

Respondents remembered and discussed their graduate education in greater detail than they did their undergraduate years. Among other issues, they talked about the value of teaching assistantships and research fellowships and the importance of laboratory experience. They also analyzed the need to choose the right institution, the right courses, the right undergraduate and graduate degrees, and the right advisors and mentors.

In graduate school, as all throughout the educational process, both students and scientists emphasized the importance of support networks. Among her many achievements, one black biophysicist and university administrator was proudest of the post baccalaureate program she developed for premed minority students. For a year, it supported financially and brought up-to-date academically minority students with demonstrated undergraduate learning skills who decided late in their college careers that they wanted to go to medical, dental, or veterinary school.

A number of the scientists had held teaching and research assistantships themselves during their student days, and some of them commented on the worth of such experience; not only because of what is learned but also because of the contacts formed. The students also generally found assistantships valuable. Said one scientist on the subject of teaching fellowships, "Teaching or tutoring is a good way to get a clearer understanding of subject matter, to understand thoroughly things you may have glossed over before. Many students who get graduate degrees in chemistry don't want teaching assistantships: they want to get right to the bench. That is unfortunate in terms of their own understanding of science."

Academic assistantships for science students often involve teaching or work in the laboratory, a process the majority of those interviewed heartily approved. An academic scientist advised that students who do not get their hands dirty, analyze the results, and make deductions, can't be scientists. "The sooner you get into the lab the better," she concluded. A government scientist echoed her advice, "Get hands-on laboratory experience as soon as you can. If you know your way around the lab, you will succeed, because you will begin to understand cells, cell mechanisms, and cell regulation. Even if you just wash test tubes in the lab, you can talk with the people who are doing science, and you will learn."

Graduate students must make many important choices. First, they must decide which field to study, where to apply, and what institution to attend. According to an academic researcher, women can use the under enrollment of students in graduate programs to their advantage. A number of respondents emphasized what they see as the importance for future career success of doing graduate work at a prestigious research university. Others advised students heading for a cross-field degree (for example, environmental science or biochemistry) to do their undergraduate preparations in a rigorous single field, and, according to one scientist, their master's level study as well.

A few of the women interviewed "fell into graduate school", not really knowing in advance it was something they wanted to do. A professor of ecology says, "I never had the goal of getting a Ph.D., but I was motivated to do what I had to do to teach as I got older." Another respondent unexpectedly earned a master's degree working in a lab while her husband finished school.

While several scientists recommended that students take as much mathematics as possible, the three women discussing kinds of advanced degrees to take reached no clear consensus on whether an M.D., a Ph.D., or both is a candidate's best course. Both degrees have their merits, they agreed. Which to take depends on one's professional and personal purpose: explained a Ph.D., "The M.D. gives you the credibility to interact with patients and design clinical studies". There is no limitation on research with an M.D., but you develop a different set of skills than you gain through a Ph.D. program. With a Ph.D., you learn how to investigate a very specific area in depth. "For unorganized people like me," quipped a holder of two degrees, "both were the answer."

The graduate students interviewed found their interest in science through paths similar to that of the established scientists: encouragement from family and gifted formal education, along with an interest in science that was never extinguished. One woman emphasized the importance of support structures, "Don't let anyone tell you that you can't do what you want to do. Try to find female mentors and peers for support. A supportive partner can help a lot as well." Like the scientists, the graduate students emphasized the importance of mathematics in entering and surviving in science; however, said one, for her "Entering wasn't the problem, staying is. The field is completely male dominated," she continued, "and I run into the consequences of that daily."

Both students and practicing scientists also face the hard reality of funding – where to get enough of it as salary or as grants. Having a relatively comfortable living arrangement and being free of financial worry is key to success in graduate school. Most universities offer students many kinds of help as they progress toward their degrees. One woman noted that she paid for graduate school with a research assistantship from her advisor. If you do need to earn a little extra money, think about what you do best and with the most ease. Always calculate the time versus income factor. Teaching – particularly as a part-time adjunct – is not always the best answer. It may be more lucrative to tutor undergraduates at your own university than to teach a course meeting 3 hours a week elsewhere.

Four of the engineers interviewed decided not to pursue graduate degrees and directly entered the workforce. Three of these women landed their first job with companies where they had previously done summer internships. One respondent, who transitioned from mechanical engineering to telecommunications, stated, "In the field of telecommunications, an advanced degree in engineering is not beneficial. There are certifications that an engineer can get in their field that weigh more heavily than a masters." She adds that a degree in business administration would be helpful for management.

8.6 The Postdoctoral Years

Seek supportive people, work on building confidence, and above all, know in your heart what you want to do and accept no substitutes.

Pamela Padilla

For Ph.D. and M.D. recipients, the next career step after graduation is to embark on a postdoctoral program. Both the scientists and the students had thoughts to share on this experience, which, said one fellow, puts one in a kind of limbo – not necessarily an unpleasant one – between the student years and full-fledged membership in the profession. Recalled one of the scientists of a former colleague, "She had been in the department as a postdoc and was well respected there as a scientist. But once she was an assistant professor, a peer, all of that changed. It seemed to me that as long as a man was validating you, things were fine, but academic science couldn't quite cope with an independent woman."

A number of the scientists provided advice on postdoctoral positions – how to win them, how long to stay in them, the status of the fellows, whether such positions are always an essential part of the professional path into science. Postdoctoral fellowships are more usual than not in the laboratory sciences and are almost always undertaken by students bound for academia, but are not so essential for those planning clinical, industrial, or mathematical careers. Explained an executive officer at the National Science Foundation.

The career path of mathematicians seldom includes the kind of postdoctoral experience you find in physics, chemistry, biological sciences, or other areas. It is more traditional to move into some form of faculty position. Those with degrees from the best schools sometimes go into the equivalent of postdoctoral positions, but they're called "research instructorships" or "research faculty" posts. They comprise half-time teaching and half-time research positions.

Jobs are often awarded through professional networks; here is another place where such contacts can be of significant value. For example, reports another federal scientist, her office awards several postdoctoral fellowships, "Since I've been here, only one person has been appointed whose major professor did not have connections here. The professors call up and say they have a student coming out who is looking for a postdoc and who is interested in this, this, and this. And somebody says, 'that sounds great,' and then the student has a job here. If your professor never makes that call, you don't have much of a chance to do a postdoc here." While there are other avenues into good postdoctoral positions – for example, one candidate wrote a letter to virtually everyone in the scientist's branch and succeeded – the government scientist warned that "You have to work a lot harder, however, if you're not connected."

Agrees another respondent, "As a postdoctoral student, you need a guardian angel to show you how to write grants, to acclimate yourself within a department, and to help you learn how to play politics." Other scientists warned upcoming women in science not to stay too long in postdoctoral positions. Expanding her advice to the career schedule as a whole, an environmental consultant noted the importance of keeping to a timetable: "An extra 6–12 months to complete their thesis research, extended duration for each postdoc, and an additional year or two before each promotion in the job. Suddenly, a woman can find herself years behind where she might otherwise have been. Then she may not be viewed as a rising young superstar, but rather as an older scientist of whom it is said, 'Well, if she's so good, why hasn't she achieved more by this age?'"

One of the postdoctoral researchers interviewed commented on how she prepared for her postdoctoral work and what she looked for in the position. She says "You either do a postdoc, or you go into industry." There are characteristics of a place and an advisor that one needs to focus on, "find someone who will be good mentor and who did research in an area [I] wanted to lean more about." If you already have a family, then things like an on-site daycare and a preschool facility are important.

Chapter 9
Moving Toward Career Success

9.1 Making the Connections

Sometimes a small effort can change someone else's career significantly for the better.
Resha Putzrath

In this chapter and the preceding Chapter 8, the perspectives of 26 women scientists, engineers, mathematicians, and administrators and 16 postdoctoral fellows and students are reflected in their views of the joys and sorrows, problems and solutions, strengths and weaknesses of a career in STEM. A number of themes were recurring in the extensive database of in-depth interviews compiled by the Association for Women in Science as part of its work on a National Science Foundation grant.

The scientific network as a whole and within its parts is a valuable resource for the exchange of ideas and information on new breakthroughs and technology. It operates within academic departments, across fields, within industry and government, and throughout the profession as a whole. Women need to break into existing networks, respondents agreed, as well as to form their own. A number of respondents reported their feelings of exclusion from certain male rituals. But said one, things are changing. She explained that the rapport with male students that she had been unable to develop in graduate school – "I had a few close friends, but I didn't go to study groups or stay up late doing problems and drinking beer" – she has now achieved with her colleagues. As her field expands, she said, "there are not many of us women, but there is very little evidence of discrimination. I think that bright newcomers are welcome."

Some of the students interviewed appropriately linked the process of participating in networks and support groups to the need for mentors. Many of the scientists discussed the value of the connections made through individual contacts with colleagues, associations like AWIS, other women's groups, and professional meetings. The umbrella network, according to several women interviewed, is feminism itself. Respondent after respondent spoke not only of her isolation as a female but also of her pleasure in seeing other women at conferences and other gatherings. (Some also spoke of the necessity of increasing the numbers of women in science, but, except in the life sciences, saw little hope of redressing the underrepresentation anytime soon.)

If women's numbers in science rise, according to a science educator, it will be when women's networks come to rival those of the "good old boys" currently in place and working so well for men. One chemist, the mother of three daughters who have become scientists, noted sadly that the numbers of women in her daughters' college and graduate classes haven't risen at all since her own school days 25–30 years earlier.

There was general agreement about the importance of attending professional meetings to make face-to-face contact with others in the field. When an opportunity arises to present a paper or a poster session, scientists agreed, it should be seized. And if the invitation doesn't happen, it should be created. Another important network that will help to insure that women speakers make their way to the podium as invited speakers – the more women organizing the conference, the more women speaking and presenting.

Most significant jobs are filled through personal contacts rather than through official notices. And, note several scientists and students, a strong network of colleagues both within and without organized groups, is a strong force for women to not only enter but also stay in the profession. Women, who are often excluded from this network, must find their way in and/or form networks of their own. Even now, for whatever reason, women's names frequently do not circulate widely.

Several student respondents expressed an interest for the role that attending scientific meetings plays in making critical professional connections. In some fields, annual conferences or workshops become invaluable networking resources. Others cited their professors as good networking tools for future research opportunities and/or jobs.

In a book committed to the value of mentors, the fact that all 42 respondents, many of them extremely busy, offered themselves as mentors to women entering science seems both generous and appropriate. In addition, an overview of the scientists' and the students' views of what effective and helpful mentors and advisors are like brings to fitting closure to this education section of *A Hand Up*.

Most of the respondents commented on what they thought were the most important characteristics for which to search in mentors or other advisors, and the role that those relationships can play in fostering connections.

Several women of color and a scientist with hearing impairments noted the importance of students seeing that a female out of the mainstream has made it in science. The latter noted that her biology sections produce more majors than those taught by the hearing professors, presumably because the students see that she has made it as a female scientist and conclude that they can too. A black science educator explains, "Students need to know that their career possibilities can be virtually endless, and that's how I can help. At engineering schools, there are few women and minorities, and it can therefore be difficult for students to find role models and mentors." An African-American graduate student says "I am thrilled when I see a black woman scientist giving a talk; then I know I can be there someday." On the other hand, another black scientist found a mentor in a white woman doctor.

A couple of scientists provided cautions. The role of mentor is not for everyone. A Hispanic pediatrician noted that competitive people make ineffective mentors, and a chemist explained that, although she had served as a mentor for many men and women in the past, at this point in a long and distinguished career she needed to pull back.

Almost all interviewees responded that they had mentors along the way. The mentors were often the influence that kept those women pursuing their field. One woman mentions that mentors do not have to be authoritative figures (e.g. professors). Mentors are people who help you get through whatever the issue is at the time; they can be friends, colleagues, professors, or strangers. One respondent developed a network of women at the community college she attended. These women work together to help younger Latina girls fulfill their educational dreams.

9.2 Leaping Barriers and Achieving Goals

> Great ideas – the original and important changes in a field – very often result from the insight of a single person.
>
> Virginia Walbot

All women interviewed were asked what (or who) was the worst hurdle for them in entering science and how they overcame this barrier. Besides the opposition from family members and from some teachers, a few women described their own lack of self-esteem as a hurdle to be crossed. Other hurdles were external. Some cited gender discrimination and sexism from both men and women; others, economic difficulties – debts, expenses, and no jobs in their area.

Some felt they had received little or no useful guidance; some, once out of the system for family or other reasons, had difficulty reentering. Others needed childcare, while three respondents had to overcome disabilities: One, her deafness; another, her blindness; a third, her dyslexia. "Introductory science courses," said one. "My religious community" said another, explaining that the order to which she formerly belonged did not welcome the idea of sisters as chemists. "Organic chemistry," laughed a third. "Then I realized that you didn't have to synthesize rubber every day; you could put those equations to more relevant use in terms of a cell."

A few overcame the barriers by avoidance – leaving the person or situation that was the problem – but the majority relied on their own sheer stubbornness to triumph over conditions adverse to their career in science.

One of the most persistent themes was, difficult though the STEM professions may still be in some ways for women, these careers offers a stunning range of job opportunities.

A respondent trained in physics and biophysics and working as a consultant in toxicology, risk assessment, and environmental policy sums up the breadth of the possibilities as follows: "When I give career day talks, I emphasize that there are many types of scientific careers. I start by saying that I'm not going to talk about familiar fields such as laboratory science, teaching, or medicine – everyone is familiar with those jobs. Instead, I focus on less well-known, more eclectic scientific careers, such as consulting or regulatory science, and how a scientific background is useful in numerous careers, such as market research or environmental work. When I was in school, I was unaware of these options. The result is that I've had four jobs that I didn't know existed before I was recruited into them."

Several of the engineers had particularly interesting career paths. One went from biomedical engineering to public policy and uses her engineering skills when working with engineers on the job. Two others transitioned from mechanical or chemical engineering to telecommunications and sales. An Asian-American biophysicist working for the federal government also emphasizes the freedom to change that science offers. She points out that, because of the rapidly changing job market, it is important to approach education as a wise consumer. Focusing on a particular area, she continues, will help to establish a place in a specific field. Nonetheless, you are never trapped. As she notes, "It's never too late to change direction, however, or to consider an unconventional career path. Ph.D. recipients who have returned to medical school can be found in research as well as private practice, and individuals who have had private medical practices can be found now doing basic research."

An industrial biotechnologist who had returned to academia agreed: "One of the important consequences of my work in industry was that I discovered a whole new range of interests. I was amazed to find that there were so many new careers that I could enjoy." It is important to take risks occasionally to continue to learn and to be challenged.

Asked what she saw as the most rewarding aspect of her career, a federal scientist laughed and said, "To do what you enjoy and get paid for it." Said another, "It's seeing students go on and do something with their lives in science." Though both spokeswomen are members of minority groups, they expressed a majority consensus. About a third of the respondents went from laboratory and/or academic science into administrative posts of various kinds. None expressed regrets, although several found the choice hard. Two offered particular insights on the similarities and differences between the two professions. According to a consultant working in several fields, who "enjoys being a manager" and who believes that more management responsibility is a virtually inevitable part of career advancement, "Outside the lab, you more frequently have an immediate effect on the manner in which science is used to solve problems. In some ways I feel I have the best of both worlds. As my current work involves a science that does not require a laboratory, it's possible to continue intellectually stimulating research while working to improve its application in a regulatory setting."

Another administrator, once a member of a medical school faculty and a university president, believes that science prepared her well for her managerial future, in which she uses "all the same values and ways of thinking." The main difference between the two approaches, she says, is between focus on the part and on the whole. "I am not the kind of person who wants to know more and more about a smaller and smaller thing. In my area.... the field was splitting rather than lumping.... I'm much more interested in how things fit together. Science will have to turn itself around one of these days and move in the other direction, but it isn't doing it at the moment."

9.3 Timing and Choices

> Get your priorities straight. Decide what you really want most.
> Maria Cordero Hardy

All respondents were asked about their family situation, whether they were married or living with a partner, and whether they had or planned to have children. (It is noteworthy that these questions might not appear in a book of interviews with male scientists, but they probably should). The balancing act between personal and professional commitments is of the first order of importance for women going into science. Neither having a career, nor being a responsible parent, nor participating in a meaningful spousal relationship is easy. The thought of doing all three successfully is at best daunting; at worst discouraging. But, although the vote was by no means unanimous, most respondents believed that women who planned carefully could, if they wished, succeed in all three roles simultaneously. But to do so, they agreed, certain hard choices were necessary.

Some put off having children until their careers were well underway. Others agreed implicitly that the importance of getting started professionally early may be overrated, and chose to bear children when they wished, regardless of where they found themselves professionally. Some downplayed their careers during the time when their children needed them most. As one of the postdoctoral fellows, who returned to school after she had helped to rear 10 children, puts it, "You have to give up something. I gave up everything except my family and work on my degree. It was three years before I read a book for fun. When everyone else went to bed, I stayed up for at least three more hours."

An Asian-American expresses what may be a majority view, "There is never a convenient time to have kids. You just decide when to do it, and you do it. Child rearing does delay you professionally. I took some time off, but I decided that slowing down was worthwhile, and I don't think it damaged my career. One postdoc down the hall had her son and was back the next week working in the lab. It took me longer. If you do choose to have a career and a family, you have to feel comfortable with your situation. If you must worry the entire day about whether your children are in an appropriate child care situation, your productivity at work will suffer. If you're convinced that you're doing the best you can with day care – and that that best is good enough – then when at work, your primary focus can be work."

Said an astrophysicist and the mother of five, were she to start over, she would be more relaxed about the opportunities that would still be available when the children were grown up. "Sometimes I got depressed," she admitted, "when other people were presenting my work, and I felt as if that were going to go on forever. As soon as I was ready, everybody was glad I was on my own."

Women made vastly different decisions about how to integrate family and career. One respondent went back to the laboratory only 3 days after giving birth to her son, and brought him to daycare when he still had his umbilical cord. She emphasizes that her son grew up to be a warm and intelligent person and wants to assure young girls that early infant childcare is an acceptable choice. Another respondent makes a commitment to spend time with her family, even "if it means a paper isn't published as fast as it could be." She chose a position at a teaching centered college, partly because she knew it would give her more time with her family. Still another choice is to forgo children all together. This was the decision of one respondent, who says "there are limited resources of time and energy and they can only be divided so much; children are not a priority for me."

Whatever the decision about the if and when of childbearing, scientists and students agreed on the need for excellent child care arrangements, one woman postponed having children until she could afford the latter. A government medical researcher, the mother of two daughters and a partner in a commuter marriage, argues persuasively that part of every job package should be day care, "Our nation needs children, but day care is often unavailable for people who need to work. Day care should be part of the working environment, like having safe water to drink."

Along with children, usually, go spouses. About three-quarters of the respondents were or had been married, and the majority of that group had chosen to have children. Virtually all these respondents praised their husbands' support and help with child rearing, and with household tasks, and (see below) with their professional needs and aspirations. One offered this tribute, as she named her husband as the greatest help in keeping her in science: "He is a man who helps people see what they can do and then sees how he can help them do it." Sharing domestic responsibilities seems to most respondents necessary and natural. Commented an academic scientist and administrator, "There is more sharing of domestic work in couples where both the men and women work. Several of my graduate students have children, and generally their husbands are supportive."

One person takes the child to day care and the other picks her up. It's not totally the woman's responsibility, "My husband has always been very supportive of what I do: He listens to me expound when I'm upset and recognizes that I may not always be home for dinner. That has been important in my success as an administrator and a faculty member because it's an irregular lifestyle. I initially made moves in support of his career, and he later made moves in support of mine. He's my best friend."

If a woman has a "significant other" who is not supportive, it can be difficult and stressful for both of them: They either get divorced, or she gives up her career. The spouses of virtually all married respondents were also employed. The demands put upon a dual-career marriage are perhaps even greater than those in the now rare (six percent) family whose stay-at-home mother takes care of her children while her husband works. Even that unrealistic paradigm by no means stays out of divorce courts. Compromise in all relationships seems essential, and various couples found various solutions to the peculiar demands of a two-career marriage, One, embraced by several respondents at one point or another, is the commuting marriage, which no one defended as ideal but which sometimes seemed a necessary solution.

Says an environmental scientist who had for a time to live in a different city from her husband, "We had big telephone bills, but we were both very involved in our research at the time, and the arrangement worked out." Says a black college administrator, on occasion also forced into a commuter marriage, "My husband has been willing to relocate with me in the past, and we have had to have a commuter marriage at times." Her phone bills also soared during those periods. Laughs an environmental consultant, "I always joke that my husband has been following me around. The truth is, however, that we've worked together to achieve a successful, two-career marriage. It's not just luck; part is what you look for in a spouse. My husband is very supportive; I sometimes tease him about being more serious about my career than I am. We both, however, have tried to keep our careers flexible and mobile. When we have opportunities for change, we examine the issues in terms of both of our careers as well as our personal lives."

To avoid the commuter marriage, compromises are sometimes but not always necessary. For example, a Hispanic scientist gave up a tenured university post to follow her husband but then found an equally good job in the new city. Several couples embarked together on the two-job search, choosing a location only if it could employ them both. If an institution really wants to hire one person, often it will use its network to find employment for his/her spouse. Although the end of the nepotism rules of the past now sometimes makes it possible for a couple to work at the same institution, finding positions for two professionals in one place is often far from easy.

One respondent balances more than a relationship with her career; she trained as an Olympic athlete while pursuing her science education. She says that "the structured schedule that she had to keep actually made me more efficient. I knew I had limited time to get things done, so I did them in that time and focused my energy. But it was exhausting." She also currently balances a long-distance relationship due to their respective careers.

9.4 Facing the Gender and Diversity Issues

> We must find new initiatives to continue the progress toward equity and be ever on guard to protect it.
>
> Anne Briscoe

Women scientists, operating in a feminist age and in a male-dominated profession that they have had to brave terrific odds to enter, tend to be sensitive in the extreme to a broad range of gender issues. Scientists and students alike were articulate and, often, angry about the many forms of sexual politics at play in scientific fields. When gender issues arise, women are nearly always at risk in some way.

Gender issues run a long gamut. On the negative side, behaviors range from the differential treatment often meted out to men and women students or professionals to active sexual harassment of women by their superiors. Also a minus is the isolation many women feel as one of the few. Being in the minority carries its own responsibilities. As one respondent says, "Every committee wants a woman to serve on it. With the lower percentage of women in each of the scientific fields, the women are being pulled thinly among a variety of committees, whereas the men might only have to serve on one."

Equally upsetting to many are those situations in which women collude with their detractors, either by encouraging flirtatious behavior in the workplace or by denying their own femininity through what has been called the Queen Bee syndrome. When racial or ethnic prejudice joins sexism, the result is even more damaging.

Some apparent gender differences carry a less negative impact. Allowing for the difficulty and danger of generalizing, a number of respondents thought they noted certain divergences in the way men and women typically assert themselves, compete, approach a task, and relate. At times, these apparent differences work against women and are therefore unfortunate. At others, feminists would assert, the female approach is better than the male's and should be adopted. At still others, the two could be complementary rather than oppositional.

One scientist, who grew up in Great Britain, notices differences in gender issues between the United States and the United Kingdom. She cites the increased emphasis that some higher education institutions in the United States are placing on addressing gender issues in academia. But she claims that in order for the problem to truly be eliminated, men in authoritative, high positions need to recognize the problem and make it a priority to eliminate it. She adds that "the younger generation of men, whose mothers were part of the feminist movement and who should have learned about gender awareness, are ironically the ones who display gender bias behavior. They need to take a 'Feminism 101' course!"

Echoing the advice of one individual who warned women to beware of belittlement masked as chivalry, a practicing psychologist defined one typical kind of gender discrimination. She explains, "If you are being discriminated against by male administrators because of your competence in a world that is unfriendly to strong women, if you are being unequally paid for equal work, for example, you are facing one kind of injustice that many, women face. What are some signs of this kind of discrimination? One clue is if you are given mysterious deadlines that don't seem necessary or real. Or deadlines may be suddenly moved forward. Or there may be comments such as, 'Don't apply; it's not time yet.' Or 'Maybe this grant shouldn't be yours but should go through someone else.' Or 'Maybe it's not time for you to be independent yet'."

This kind of differential treatment of men and women can take place in the laboratory and other workplaces, in academic settings, at scientific conferences, or in any professional area. And it can happen to women of any rank, from students to top-level researchers and administrators.

An academic chemist notes how surprised she was when she first started out as a professor, "At first I didn't think I would face any different hurdles than others. I grew up in a very egalitarian family and went through grad school and my postdoc with very supportive advisors. It wasn't until I started my first faculty position that my eyes were opened. I then became the victim of petty discrimination. The biggest problem with this is that it takes up a lot of time and produces a lot of stress. For example, I have struggled to be allowed to teach a special topics course. I've been at this university for 18 years, and several junior male faculty have taught their courses multiple times. I finally get to teach my course this year, but it's been a long and draining fight."

According to an academic marine biologist, this kind of prejudice, which has long been a factor, has a continued presence in the academy, "It's just underground. There is kind of a 'snicker-snicker' pretense that administrators practice – they have to act as if they were open and giving women opportunities. This is a generalization, and perhaps it's unfair, but some male researchers use women graduate students as technicians, no matter what the detriment to the women's careers. That's changing, but what I often see is women getting stuck at low academic levels. There is a glass ceiling, and it needs to be smashed. It's going to take a lot of persistence, because women are recognized as good assistant professors and laboratory technicians, but they don't get promoted as readily as their male counterparts. There aren't very many women senior professors. There aren't many women in most science departments in most universities, and that is the result of sheer prejudice. Jobs are scarce now; therefore, any excuse is used to cut someone out, and a 'different' person is the one they cut."

And, according to an academic chemist, this approach is neither new nor confined to faculty. She remembers her graduate school days in the 1960 when the handful of women chemistry graduate students taught the women undergraduates – mostly home economics majors – especially tailored laboratory sections, and "the male graduate students taught everyone else." While she admitted a better climate now, she remembers that "that kind of treatment was everywhere then, and no one thought it was wrong."

Another returning student remembers being told she was "too old" to enter graduate school. Though her interrogators were overruled and other faculty were "generous and supportive, it wasn't fun at the time." It was especially discouraging for this woman, who had left a job "where things were not much better" to pursue her Ph.D. A government scientist sees "two pathways: one for men and one for women. It's subtle, but women don't get the salaries or the promotions." She continues, striking a familiar coping note, "With wisdom and experience, you learn to fight for certain things, let go of others, and not get discouraged," concluding, "I think women have to work 10 times harder than men to get promoted."

A number of women brought up their anger at the sexist jokes they were expected to endure cheerfully. A federal grants officer remembers during graduate school in the 1960 that it was considered "cute, indeed complimentary, to sexually harass a woman student. This was something a professor would boast about, and the department would make jokes about." At least now, reports a doctoral candidate, although there are still professors who tell such jokes, "Those jokes fall flat when more than half the students in the class are women."

One woman engineer turned the joke back on her boss. She said she found her first real experience of gender discrimination when she married, "Everyone except my immediate supervisor assumed that I'd quit work and forget about a career. Once others realized that I was serious about my career and that my husband was supportive, things slowly changed. Even 10 years later, however, people worried about promoting me because I might get pregnant. I coped with this problem by forcing a meeting with the manager. After discussing technical matters, I ended the issue forever. I said as I left his office, 'I'll make a deal with you. In the future, I won't ask about your sex life if you don't ask about mine'. Today, he tells the story with a touch of humor and adds that it was the most consciousness-raising experience of his life".

Not everyone, unfortunately, can laugh off and turn away their tormentors so easily. And the result of being treated as a token or a joke or an outsider can lead to the profound isolation reported by a number of respondents. One woman notes, that although she is often one of but a few women at meetings, the situation is improving. "The problems, however, remain," she points out. "Many difficulties formerly overt are now merely buried." Another remembers meeting organizers trying to register her as a spouse rather than a participant. And a third respondent, a woman astronomer expresses what amounts to a consensus statement on the feelings of exclusion suffered by many, perhaps most, women in science, "You show up at a meeting, and you're the only woman there. Every word you utter is remembered in a way that a young man's isn't. I survived. The real tragedy is the hundreds of women who would have made outstanding astronomers but never made it because they were discouraged all along. Academic and professional circles are still male dominated. You will go through 3 or 4 days of meetings and never once hear the word 'her' used. Every scientist is 'he.' Maybe this is just language, but after a few days, it becomes very annoying. I write a letter every year to the National Academy of Sciences objecting to something they have just sent me in which all the personal pronouns are masculine. I think we're kidding ourselves if we don't realize that science is still a male-dominated profession, and some of the males enjoy this dominance."

Isolation is bad enough, but even worse is the active sexual harassment respondents at all levels of academia, business, and the federal government report. One interviewee had to fend off "blatant sexual advances" from a man who would later turn out to be important in her field. Another woman, speaking for a number of mature women scientists, said, "Of course," she had experienced gender discrimination. But, she went on, "instead of recounting experiences, I would like to broaden the question to discuss how to respond to such situations. For example, what do you do the first time you see a Playboy slide in a lecture? You behave differently than you do the fifth time such an image intrudes into a professional setting. I would advise young scientists to talk with others to see how they have handled specific situations and think about what response is comfortable for them for each case. Being caught off-guard can produce regrettable responses. Sometimes, a meaningful glare is sufficient; other times, stronger action is appropriate. In some circumstances, it is best to change things quietly; other times, vocal support is needed. Find out what has worked for other people and then adapt those ideas for your own use. I try to focus on what result I am trying to achieve rather than why the action has upset me; this approach allows me to stay more objective."

At times, however, the situation becomes such that legal action may appear the only solution. In a case when sexism took a particularly ugly turn, the harassers were not men but other women colluding with lascivious superiors, a government psychologist decided that she bad "had enough," and she sued on the grounds that she should not have to work in a "hostile sexual environment." Though her action was absolutely necessary, a lawsuit is a draining procedure, and she had some other advice, "I encourage speaking out against harassment, but you must be aware of the possible ramifications. You may be unpopular with your colleagues, and roadblocks may be placed in your way. Before you accept a position, make sure that it is part of a cultural environment where you feel comfortable and where YOU think that your views and research would be encouraged and rewarded." In any case, it is good to talk to someone who has seen discrimination before and can make suggestions for how to handle those very complicated matters.

Several other respondents discussed certain kinds of female anti-feminism. Some women, notes a biology professor, may infight among each other. "This sets us back," she said, "because the pie is so small. We are all fighting for a bit of it instead of trying to make the pie bigger." Also to be avoided is the Queen Bee syndrome, whose participants advance by being like men. Such women, points out a science educator, are not particularly nurturing to women. Concurred another, "Many years ago, I listened to women who were supposedly successes and felt that many had paid a dreadful personal price for their professional achievements, a price that men need not pay. Women should be able to choose whether or not to buy into the male game. A black and a Hispanic scientist concurred on the particularly poisonous mixture of sexism and racism. One fights it as follows: I have never been able to separate the effects of being black from the effects of being a woman. I was aware of racism, of the feeling you can't afford to fail because others will have a problem because of your failure. I also knew that there were boys who resented girls who did better than boys. Each individual must make her place in the world. It helps to know that paths have been made. People have survived, and you can also. There are people who are supportive. I wish there were better ways that women could support one another, so each woman could do what was right for her."

Also aware of what she called "a wave of racism within the power structure," adding that her colleagues are "not necessarily guilty" of it, the Hispanic pediatrician explained that "People's ideology somehow makes them distort others' realities, and people see you not as you are but as worse (or better) than you are because of their prejudices about you."

Many of the minority women interviewed felt they had experienced more racial discrimination than gender discrimination. An African-American chemist says, "When I was starting my professional career, I applied to three pharmaceutical companies. I was offered positions at two companies, but the third company took one look at me and said they were no longer hiring. One of the reasons I decided to get a M.S. degree was that I knew I would need the M.S. to get the same types of jobs a white person would get with a B.S." However, she goes on to say that the situation in the pharmaceutical industry is now much better. She has even participated in company projects to encourage more African-American students to apply for employment.

Several respondents tendered opinions on areas where men's and women's approaches appeared often to differ, only sometimes to the detriment of women. There are many ways of knowing. The two sexes tend to approach the same problem in different ways and communicate their ideas differently, according to one scientist. Women are more open and expressive, according to another. "I think that men may look at things with more tunnel-vision than women," said a third. While it is important that women not be so retiring that they are ignored, a pattern about which several respondents worried, they should not become like certain aggressive men or like Queen Bees either. Instead, noted one scientist, "There's another factor we're only beginning to recognize: Women work differently. Women tend to run very open shops. Information is accessible – everybody knows what's going on. Men see information as power, and they share it only grudgingly. These factors must be better understood, for they are critical in getting women into positions of power."

But there is no clear consensus on gender differences. To conclude with the opinions of three respected women scientists on the matter:

- Stereotypically, men tend to approach tasks in a more concrete, structured way, while women tend to see the totality of the task and do not necessarily choose to do step one first. In this respect, my husband and I are opposites, but not in the way you might expect. He's far more intuitive, and, because of my training, I approach a problem in a much more structured way.
- I can't make a generalization about men and women. Even when I talk of a man and a woman, the matter is complex: While my husband tends to be interested in theoretical matters and I in practical aspects, my oldest daughter also tends to the theoretical.
- I think women approach things with a more intuitive sense; they go with their gut feeling, pursuing a lead whether or not there is hard data to back it up. Men tend to want hard evidence before they go off in a particular research direction.

9.5 Staying the Course

> Women entering STEM jobs are seeing others already in the workplace; and they recognize many of their established colleagues as well-rounded individuals who enjoy and tend both their careers and their families.
>
> Eleanor L. Babco

During the course of the study reported in this chapter and in Chapter 8, women have proffered advice on many professional and personal subjects. They have suggested ways to smooth one's educational path and to move into a job. The next problem is how, if she wishes, the new employee can become a veteran. Respondents share their advice in terms of what was the most helpful in keeping them in science or mathematics once they had entered the field. Scientists and students usually credited several sustainers, but far and away the greatest gratitude went to their mentors and teachers at all post elementary levels. They also cited their own determination and interest in their work and their family's support.

A chemistry graduate student offered an excellent suggestion for how to keep girls in science, "What helps to keep people in the field is direct positive feedback. People rarely say 'You are an asset to science, you should stay in the field.' I only remember hearing that once. Many women have less of an innate sense of self-confidence, so they need more external encouragement."

Several women pointed out that changes to science education may help young girls, and all children, maintain their early curiosity about science. A chemist who transitioned to science education and consulting emphasizes the need for motivated science teachers, "People say all children can learn, but they don't actually believe it. Teachers pick out their favorites and don't encourage the others. Kids really need to be challenged. It won't be easy – kids come to school with attitudes – but you need to listen to them and encourage them and make them work. Teachers need to believe in the kids, whether there are 2 students in the class or 90."

Many of the student interviewees, still participating in the education process themselves, were particularly animated on the topic of science education. A chemistry student remarks, "Any kid is interested in doing things, so early education should emphasize learning by doing." An undergraduate studying engineering explains a useful technique used in her science and mathematics high school, "We used an iterative approach to learning called problem-based learning. You look at a problem and ask yourself, 'what do I want to know about this?' Then you go do research, in the library or on the web. You come back to the problem and say, 'what do I know about this problem now and what do I still need to know?' I think as long as you can ask questions, you can come to an understanding of the fundamental ideas of science. This technique could be brought to any class and it would help students become interested in science."

Some students already have plans for the courses they will teach later on in their career. States one, "The course I assisted in at the university was very oriented to real world problems, and I think the students liked that. The professor would teach radiation in the context of using radiation to kill tumors. That is how I will teach when I run my own courses some day."

A majority of respondents offered "advice to a young woman thinking of embarking on a scientific career." Like the women themselves, the responses are richly individual, useful, quirky, and to the point. For example,

- Gravitate toward people who have bad attitudes. That means they are thinking – Darwin, Freud, the Curies, and Pasteur all were asking disturbing questions.
- There's no use saying that you should have done something sooner, or that it's too late now.
- Do what you love. If something interests you, you will succeed in doing it.
- Get involved in lots of outside experiences, whether they are research, summer internships, co-ops, jobs.

Nearly a third of the scientists and students counseled persistence in the face of discouragement. "Stay the course," said one, speaking for many. "Don't take no for an answer," agrees another, "and remember: If somebody else can do it, so can you."

Chapter 10
Voices of Experience

10.1 Women Speakers: Make the Most of Your Moment

Heidi B. Hammel, PhD, is a senior research scientist at the Space Science Institute in Boulder, Colorado. Her main areas of interest are ground- and space-based astronomical observations of outer planets' atmospheres and satellites at visible and near infrared wavelengths. She has won many awards for her education and public outreach work. She lives in Ridgefield, Connecticut with her husband and three children.

It is the afternoon of the third day of the meeting. Hundreds of talks have been given already. The lecture hall is dark; the only sound is the droning voice of the man giving the talk. He's the fourth speaker in a row, and they are all starting to sound alike. You fall into a semiconscious daze.

Applause momentarily jars you to consciousness, but the voice of the male moderator announcing the next talk quickly lulls you back to somnolence. Suddenly, a woman's voice starts speaking clearly and strongly. The sound of her voice – so much higher in pitch than that of the previous speaker – catches your interest.

This moment is critical for the woman speaker. The sound of her voice is unusual and catches the attention of the audience. By virtue of women's limited representation in science, such interest-catching opportunities occur with some frequency at scientific meetings. Women in science, especially female graduate students giving their first public presentation, need to be aware of the importance of this recognition.

A male colleague of mine, speaking about a woman student, once told me, "Boy, she really blew it at that last conference. It was her first talk, and she gave a horrible presentation. Since there were only a few women there, everyone knew exactly who she was and for whom she was working, and her performance didn't do her reputation any good, that's for sure." Another colleague put it this way; "A young female speaker just can't just fade into the yet-another-bearded-male-graduate-student horde."

This is reality – women speakers can make a striking impression. Instead of bemoaning this fact, the woman speaker should turn it to her advantage. Below are some tips for tightening a talk. While geared toward short presentations at scientific conferences, these principles can (and should) be applied to longer colloquia and speeches as well.

10.1.1 Memorize Your Introduction and Conclusion

The introductory and concluding sentences are, by far, the most important ones of your entire presentation. If you prepare nothing else, at least prepare these sentences. "You never get a second chance to make a first impression," warns an old TV commercial. Make that impression a good one. One woman I know always had a tough time getting her talks started. Once she was a few minutes into them, she relaxed, but by then she had already lost her audience. After she learned to memorize her first five sentences, she was able to stand up, take a deep breath, and start talking loudly, clearly, and without hesitation.

If you lose the audience's attention during your talk (and this happens to the best of us), the words "to summarize" or "to conclude" will usually bring it back. Make sure you exploit your advantage by providing a concise, clear conclusion or summary. Don't babble for another 5 minutes and fade out with something timid, such as "Well, I guess that's all I have to say." Such an anticlimax leaves listeners feeling cheated, frustrated, even annoyed. Leave them instead with a pithy summary.

I used to literally write out, on paper, my first and last couple of sentences. I rarely read them – they were usually fixed in mind as I think through exactly what I want to say. But, once in a while, I had a train wreck, and having my well-worded conclusion written in front of me salvaged a potentially gruesome experience.

10.1.2 Talk to Your Audience

Face your audience: Do not talk to a screen or blackboard. If you are unamplified, your voice will disappear the moment you turn (that's why actors orient their bodies toward the audience during stage performances). If you are speaking without a microphone, you should be especially aware of volume, always speaking so that the people in the back of the room can hear you clearly and distinctly. If you are unsure, ask: "Can you folks in the back hear me?" But, even if you are amplified, don't talk to the screen. Both regular mikes and clip-on types lose your voice the minute you turn your head.

In the best of all possible worlds, you won't need to point out specifics on your slides because they will be clear and self-evident (see below). In reality, though, it is sometimes necessary to use the pointer. If you must point to the screen, try to pause, turn and indicate, revolve back to your audience, and continue talking. Doing this gracefully – while keeping the "flow" of the talk going – is not easy. That's why you need to practice. Pay attention to how other speakers – especially good ones – handle this.

10.1.3 Watch the Clock

If your paper is too long, cut it. The moderator usually has a timer, but s/he will only signal when your time is running out. It's much better to keep track yourself, so you can adjust if you spend too much time early on. I often take my watch off and set it on the podium, where I can see it with a glance. That way, I'm not constantly looking at my wrist (a distraction for the audience). Practice your talk ahead of time to check the length.

If you still run out of time, jump immediately to your prepared concluding sentences and end gracefully as you planned. The worst thing to do is to speed up. Resist the temptation! It's better to leave something out. (You may be able to bring it up during questions afterward.) One way to control time variables: Prepare your talk in "modules," and be prepared to drop several near the end, if necessary.

Never drop your conclusions, however, and make sure that your final sentences are appropriate if you do have to end earlier than you expected.

10.1.4 Use Visual Aids Carefully

I give most of my talks as electronic presentations now. If that is not an option, I highly recommend slides. I prefer both of these options rather than viewgraphs (also called overheads, films, or cells). Electronic presentations and slides force you to organize your talk ahead of time, if only because it takes time to create them. With overheads, it is too easy to wait until the last moment and then scrawl them out. Overheads are also messier to handle than slides or a portable computer. Electronic presentations can permit last-minute changes, but at least the modifications will look neat and be consistent with the rest of the presentation. If you do plan an electronic presentation, bring along a back-up in viewgraph or slide form.

If you must use viewgraphs, have someone else change them during the presentation. (You will have plenty to concentrate on during the talk.) Also, when someone else is changing your viewgraphs, you have to be well prepared: you won't be able to rearrange them during the talk. If you must change them yourself, be careful. Nervous hands can knock over a pile of viewgraphs. Whatever visual aids you use must be carefully checked before use.

Look at the screen from the back of a room comparable to the size of the one in which you are talking. Is the text legible from a far corner? I have been at talks where viewgraphs were unreadable even from the front row. Such "illustrations" are worse than none at all; not only do they add nothing, they actually distract the audience. At another presentation, the slides were entirely unintelligible. The speaker apologized – "Well, this slide would show the X instrument, but it's kind of old, so you really can't see the apparatus. This next slide would show our setup if it weren't so dark." The slides were a waste of time and an impediment to understanding. If the presenter didn't care enough to get decent pictures of his work, why should his audience care about what he was describing?

10.1.5 Practice, Practice, Practice

First, practice the talk itself, checking its content. Volunteer to give it as a lunch talk. If several of you are going to a conference, arrange a pre-conference mini-symposium. Or, even better, do both. Ask for constructive criticism after your first attempt, on both the content and on your presentation. Apply what you learn to your second try, and then solicit comments after that try, too. If you are a mentor, strongly suggest that your graduate students give practice talks before they go to a conference.

Second, practice the actual physical presentation of the talk. When you get to the conference, find out where you are speaking, and go look at the room. Get the feel of the room. Try using the pointer. Pretend to talk in the microphone. Practice adjusting the mike to your height. Gauge the size of the room, thinking about the person sitting in the last row (what can s/he see and hear?). Put up a view graph or a slide, or hook up your computer. I usually do this the morning before I give my talk or during a coffee break before the session – early enough so that people aren't there yet. (I feel silly sometimes, but I'm more comfortable later on during the real thing.)

Suddenly, a woman's voice starts speaking clearly and strongly. The sound of her voice – so much higher in pitch than that of the previous speaker – catches your interest. She speaks precisely, introduces her material neatly, and sets it carefully into her talk's context. Her slides are readable, colorful, and concise. You feel as if she's talking to you personally; she even seems to look straight at you for a while. She doesn't seem rushed, and she reaches a logical conclusion, which sums up her argument in a few words. You will be able to explain correctly the gist of her remarks to a friend later.

Next time, that speaker is going to be you!

10.2 Things your Professor Should Have Told You

Alice S. Huang, Ph.D., is Senior Faculty Associate in Biology at the California Institute of Technology. She was previously Professor of Microbiology and Molecular Genetics at Harvard Medical School, and subsequently Dean for Science at New York University. She has served on the Boards of the Rockefeller

Foundation, Waksman Foundation for Microbiology, and Public Agenda. She currently consults on science policy for government agencies in Singapore, Taiwan, and China as well as for the US National Aeronautics & Space Agency, and the California Council on Science and Technology.

As recent as 30 years ago, it was not uncommon for male professors to ask female graduate students "Why do you want to go into science when you can be at home raising beautiful babies?"

A lot has changed since then. Over 60 percent of married women and 78 percent of women with children now work outside the home. Many jobs previously thought to be unsuitable for women are now available to them. Women visibly participate in every part of society. Yet recent surveys show some disturbing trends. A larger percentage of women entered the science professions in the 1970s than in the 1990 s. Barriers to women's career advancement, although they are more subtle, still exist. How can we remove those barriers? How can we encourage young women to enter the sciences and become successful science professionals?

10.2.1 Gaining Opportunity, Equality, and Power

In 1995, then Secretary of State Madeleine Albright said that women will contribute fully when they have opportunity, equality, and power. In reviewing changes over the past 30 years, we can say that women have largely gained equal access to opportunity. But equality and power still elude many. Without full equality and effective power, we cannot contribute fully to society. More importantly, we cannot better our own lives or those of our daughters. To gain full equality, we must gain power. Therefore, power should become our next focus.

How do we gain power? It is not a disgrace to want power and to wield power, especially when it is for a common good. We tend to forget that having power, being in control, can be exhilarating. To gain it, we can try to shame others into giving up power, but those men who have it are not likely to give it up voluntarily. We can lobby government agencies to pass laws that will protect and help women, as we have done effectively in the past. However, such laws cannot affect a transfer of power. We can ask both public and private funding agencies to provide grants as incentives or rewards for hiring and promoting women into positions of power, as has been done with limited success. We can provide a list of "best practices" to help institutions attract, retain, and promote more women. But all these efforts depend on persuasion, and affecting real change is likely to take decades.

Because external power is not readily within the reach of many women, we need to focus on self-empowerment. This is within our control but, unfortunately, it is not often done. Much can be gained if we practice self-empowerment as well as empowerment of one another. Self-empowerment means celebrating and supporting women as well as sharing our experiences and educating each other about what leads to success. This empowers each other and ourselves.

Let me share with you what I have learned in my career as an academic scientist and university administrator about power and empowerment. It is important to understand power and how to gain power in our own right. My examples are from the biomedical sciences because that is what I know best. Nevertheless, the ideas are applicable to women – and some men – in other areas of science and beyond academia.

10.2.2 Learning the Academic Structure

My advice begins with understanding the structure and culture of the working environment. In higher education, as in many professions, individuals pass through specific gates in the natural progression of careers. Each of these gates is marked by a title change and an increase in salary corresponding to years of experience. We are familiar with the academic ladder beginning with postdoctoral fellow, promotion to assistant professor, to associate professor, and so on. If one chooses not to follow this well-defined path, there are other routes to take, but it means getting off the academic ladder.

Defining a new career structure can be rewarding, but often getting off the ladder results in difficulties and disillusionment. For example, a research associate position is commonly sold as a job with less stress and more freedom to pursue research. In truth, there may be less stress, but freedom is illusory. Proceeding along this path provides a chance to gain research experience but not commensurate increases in salary or public recognition. As the years go by, increased professional isolation takes its toll; despite maturity and experience, reversing this projection and getting back on the academic ladder is extremely difficult.

Another reason to leave the usual career path is financial. Sometimes, there may be an offer to be a research associate or a laboratory director. Taking a lucrative but subordinate position with a faculty member at the university can be a compelling incentive to jump off the academic ladder. Often this move is based on promises of increased responsibility. However, the once lucrative salary quickly reaches a ceiling. Further advancement is limited and job stability depends on the tenure of the faculty boss at the institution. Should the faculty member not gain tenure or decide to move to another institution, it may be difficult to gain an equivalent well-paid position with another faculty member.

Leaving academia and joining another structured environment, such as the biotechnology industry, offers financial rewards and unusual challenges. However, once this route is taken, proprietary information may limit publication and getting back on the academic ladder becomes more difficult if not impossible. If one succeeds in returning to the university, however, there is a substantial reduction in income.

Although there may be good reasons for taking these different paths, it is important to be fully aware of the consequences of such choices, especially when they are made early in one's career. In the academic culture, falling off the academic ladder means leaving the usual path to advancement, security, and recognition. More importantly, these other routes do not lead to power within the academic structure.

10.2.3 Starting off Right

Once the structure of the organization is understood, it is necessary to be successful within that organization. To do so means fulfilling the expectations of the organization. The first independent position, usually as an assistant professor, is very demanding. It becomes necessary to teach, attract, and mentor trainees; to set up a laboratory; and to organize independent research. This is a crucial time for concentrating on one's career. Personal issues that intrude at this time may be detrimental.

Unlike students and postdoctoral fellows, an assistant professor cannot accomplish her responsibilities alone. No individual, no matter how capable, can do all that is needed at this stage as a loner. Team sports teach about cooperation and interdependence – use that knowledge. Building support, seeking out advisors, and forming meaningful relationships with colleagues are essential at this time in your career. There are many ways to accomplish this.

First, other women, especially secretaries and technicians, are there to offer support and they can be tremendously helpful, especially if they think they are respected in return and are appreciated for their contributions. Delegate, delegate, delegate! Delegating routine, time-consuming tasks is necessary, no matter how well or easily you can do them yourself. When I was an assistant professor, the all-female typing pool supported me and worked on my grants and manuscripts first. Through them, I learned about the subtle, nontransparent value system at Harvard Medical School. When I discovered that my starting salary was lower than that of men hired at about the same time, these long-time staffers helped me to negotiate a salary adjustment quietly and behind the scenes, so that I did not embarrass my supervisor. They saved me from appearing to be strident and demanding.

Support can come from peers as well. All too often, we compete against other assistant professors, because in some institutions only a few survive the promotion process. A way around this competition is to seek out those at the same stage in other departments or institutions, particularly women or individuals who share similar scientific interests. We all need reality checks with peers so that we can judge whether a situation that is new to us is unusual or expected. Peer support can also provide relief in fulfilling obligations during emergencies: when I could not give a lecture, a fellow faculty member from another institution filled in for me. The students welcomed this change and my supervisors were none the wiser about my dereliction.

Support from mentors and senior professors must be cultivated, especially support of thesis and postdoctoral advisors. I am surprised at how many trainees burn those bridges unaware that future employers and promotion committees will return to old advisors and department chairs for recommendations. It is not enough, however, to maintain cordial contact with these mentors. Seek out scientific leaders and those whom you respect in your chosen field. Make sure your department chair and your dean know something about your work. It never hurts to send a packet of your reprints to all these individuals. Even better, send them preprints because those are more likely to be read. Ask them for advice and help when you need them. Most senior faculty are flattered when asked and are more than willing to help.

10.2.4 Avoid a Common Pitfall

It is likely that male professors will become your mentors, so it is important to be aware that the ugly head of sexual tensions may turn up when you least expect it. Such topics are usually not discussed because they are difficult. A good mentor is likely to become a friend. Be business-like and professional at all times. Sometimes it will be up to you to defuse tensions and make the men around you feel comfortable. Remember that you can be friends with your mentor's wife and show that you are not a threat. Jealousy on her part will inhibit mentoring by her husband. Any sexual innuendo can diminish your credibility and ruin your career. Be very careful! These are sensitive issues and it is better to be aware of them than to turn a blind eye.

10.2.5 A Word about Extracurricular Activities

Assistant professors before tenure need to use extracurricular time judiciously. Do not volunteer to be on any more committees than you have to. Gauge the value of the committee in terms of career networking and advancement. Committees to gather data on other women, to help run joint service centers like animal facilities, or to advise graduate students are often offered to women faculty. These are time-consuming and should be avoided if possible. It may be difficult to say no to some of these committees, but at this point in your career it is necessary to stay focused on the academic ladder. Pick visible, leadership roles within the institution as well as those that will enhance your national scientific reputation.

Join professional organizations and volunteer for leadership positions in those organizations. Professional gatherings provide a wealth of informal information beyond the scientific exchanges and permit you to compare your situation with many others. Information gleaned at such meetings will make you more effective on the job, and colleagues you meet may become part of your national support team.

10.2.6 Be a Good Mentor

Learn how to be an effective mentor yourself. Do not be more critical of female students than of male students. Do not be a perfectionist; many women scientists set extraordinarily high standards for themselves and for others. Promoting the best in your students will ensure a stream of trainees. However, being critical of their every effort will frustrate and turn off students. At a time when students are still unsure of their own capabilities and prioritizing their own commitments, particularly young women, they need all the encouragement they can get in order to stay in the race. They do not need what is called "tough love." Learn to compliment your trainees and junior women faculty. Compliment them not only in their presence but also in their absence. You will empower them by these actions. All too often, women faculty and students do not receive the positive feedback and recognition they deserve.

10.2.7 Make Your Work Visible, Known, and Valuable

Do not imagine that by simply working hard and being an excellent scientist you will be recognized and promoted automatically. Publishing is essential. So do not delay publication waiting for that piece of data that will make it more perfect or that will make a more complete story. Your work is your life's blood and communicating it whenever and wherever you have the chance will advance your career.

Even that is usually not enough. Helping someone else get a job done may be gratifying, but unless you lay some claim for what you have done, the credit will go to others. Some self-promotion is necessary. Seek credit. Make oral or written annual reports letting department chairs or deans know about your accomplishments and awards. Ask for promotions and salary increases. Do not expect them to come your way unless your organization has a transparent policy applied evenly to everyone. Notify the school paper or magazine when an award comes your way so that it will be properly publicized. Ask supportive colleagues to make award nominations or to suggest you for better positions.

Finally, do not ignore the finances of everything you do. Money talks. Bringing in an extra grant or an umbrella grant will empower you. Obtain a fair salary that reflects your importance in the organization. If your salary provides extra income, try contributing to your own institution or to philanthropy and see the added benefits such actions will bring. In fact, understanding and using the power of money is one of the first steps to rising into powerful management positions.

10.2.8 Once in Power...

Although some power will accrue at every level in academia, the power to change institutions really exists at the full professorial or administrative positions. There is a caveat. Polly Bunting, a past president of Radcliffe, said "Once you are in a position of power do not forget that you are still a woman." She was afraid that in climbing the academic ladder women would adopt the masculine culture and identify only with the male power structure.

I advise that early in your career you need to focus primarily on the imperatives of the academic ladder, but once in power there are many things a woman can do to help other women. Besides hiring and promoting more women, the lives of women faculty can be empowered by powerful individuals acting in the following ways:

- Review compensations, start-up packages, office and laboratory spaces, and access to institutional resources every now and then to ensure equity between male and female faculty.
- Provide discretionary dollars to faculty from an institutional source when special circumstances dictate the need.
- Introduce faculty to lucrative consulting activities or other extramural opportunities as appropriate.
- Make women faculty aware of such opportunities and how to qualify for them.
- Avoid overloading women faculty with teaching and committee responsibilities.
- Provide effective mentoring and timely reviews.
- Nominate women for awards and other kinds of recognition.
- Develop complete intolerance for the casual discrediting or minimizing of women's contributions and accomplishments.
- Make sure that the bar is not set higher for women than for men.
- Cooperate with other institutions to provide jobs for accompanying spouses.
- Provide a menu of benefits for all.
- Provide well-run, inexpensive daycare centers, as well as emergency childcare.

Many of these recommendations are found in recent national reports on the status of women and resonate with women who have long been in the academy. Some institutions have already incorporated some of these "best practices" and have found that doing so did not bankrupt the institution. Practicing all these recommendations will go a long way in improving the climate in academia for women and will help retain more women for the long haul. Only when more women gain and use power can we bring about real and lasting change.

10.3 Applying for Fellowships or Research Grants

Barbara Filner, Ph.D., served several terms as the President of the Association for Women in Science Educational Foundation. Trained as a plant physiologist and biologist, she switched into science policy at the Institute of Medicine and subsequently was a grants program administrator at the Howard Hughes Medical

Institute. She was a longstanding advocate for women in science, including election as national president of the Association for Women in Science. During her lifetime, she was a strong mentor, role model, and advisor to many women in science.

With few exceptions, a scientist's professional life from the very earliest stages is filled with grant applications. Whether as a graduate student seeking pre- or postdoctoral fellowship support or as a faculty member seeking support for her research, a researcher's chances of successful applications can be enhanced by following a few simple guidelines.

10.3.1 Graduate School and Postdoctoral Fellowships

- Identify Several Possible Fellowship Sources.

There are numerous funding programs available, from government agencies, foundations, professional societies, industry, and other public and private sources. Some are highly competitive, others less so. Try to identify as many sources as possible that might meet your needs, and consider applications to several programs. Use Web searches to find some of these programs. Print publications also may be helpful – your school's library should be able to help you find the right listings. Also seek advice from faculty members and financial aid or graduate school offices.

- Request Current Program Guidelines.

Once you find the name of an agency that provides relevant fellowships, write or call to request program guidelines, or go to the funder's Web site. The eligibility, forms, and deadlines may change from year to year, and you should be sure to have the most recent information. You don't want to miss a deadline, and you don't want to waste time applying for a program for which you are not eligible.

- Contact the Funding Agency.

Grant programs each have their own style. Some are quite willing to provide supplementary information – others will simply refer you to the published guidelines. Ask for a list of previous awards – the topics may suggest whether or not your area will be competitive. Priorities can and do change, however, so don't dismiss any program on this basis alone. Ask to speak with a member of the program staff about your particular field of research. You might also ask about award rates – what percent of applicants are funded in your field? If you are considering several possible applications, this information might help you decide where to focus your limited time. Some agencies do not make this information available, but it doesn't hurt to ask.

- Plan Ahead to Arrange Reference Letters.

Two critical steps lead to good reference letters. First, be sure faculty members know your abilities; second, be as certain as possible that the requested letters will be positive. As a student, one of your most serious responsibilities is to have several professors know you well enough to write strong reference letters on your behalf. It's your job to be sure these professors know you and your strengths in depth – beyond your doing well on tests. For science, research is the heart of the matter, and several professors must be able to say convincingly that you think rigorously and creatively and that you know what research is. Seek opportunities to show your capabilities to them, including volunteering or working for pay on research projects. Also, seek opportunities to discuss ongoing research projects with professors in your department.

Ask for reference letters in such a way that someone who is reluctant will be able to decline gracefully. You might believe someone should write a good letter, but will s/he? I know situations in which students completely misjudged the evaluation they received. To avoid this pitfall, don't simply ask "Will you provide a reference for me?" Rather, frame your request along these lines – do you think you know me well enough to write a good letter of reference?" This strategy, of course, depends upon having several possible references – hence the advice above. Giving your reference a copy of your resume also is helpful.

- Provide Both Context and Detail in the Research Plan.

Postdoctoral applications will ask for a research plan. Graduate study applications may require a research plan and/or a program of study. If the application asks for the latter, be sure to provide a research plan as part of it – characterize your plan as representative of the approach to the kind of scientific issues you wish to address and that you are preparing yourself to investigate. Clearly state and develop the research project you wish to under take, while placing it into a broader context.

Right at the beginning, you should articulate the fundamental question you are asking and why it is of interest. Frame questions that probe mechanisms. Do not characterize your research simply as "comparing" or "describing." Then, present your specific project and explain how it addresses the fundamental question you described. If you are using a model system, discuss the advantages (and any drawbacks) it offers. Don't get bogged down in technical detail, but present enough information to convince readers that your project is feasible and that you have sufficient expertise to carry it out successfully. Reviewers also like to see that you have anticipated possible problems in your plan (and how you would deal with them). If possible, close with a brief paragraph that discusses the next steps after completion of your project: Where is it going?

Reviewers want to see that you can think rigorously and scientifically and that you ask good questions. They will adjust their expectations to the level of training involved – with increasing sophistication and detail expected as you progress from student, to postdoctoral fellow, to independent scientist.

- Relate the Research Plan/Plan of Study to Your Career Goals.

Fellowship applications often include a section on your long-term goals. You should present fully formulated goals (not "I want to do research"). Reviewers also want to see that you are thoughtful (and astute) about the training you hope to receive. Explain how the proposed research or plan of study relates to your goals. Indicate why you have selected this project, mentor, and/or department in the context of your overall scientific aims.

- Seek Review and Comment on Your Application.

Experience teaches a reasonable balance between context and detail, and you will not necessarily make the right judgments in your first applications. It is vital that you get feedback from people with experience of reviewers' expectations.

After you have drafted your research plan, ask your mentor and/or another professor to read your plan and critique it. Your research plan will be strengthened, based on your responses to their questions. Perhaps the review will also occasion a serious discussion with the professor, who will then be better able to write an informed reference letter.

You might ask some other students to read the application as well. Make it clear that you want honest comments that will help you improve the application. Be prepared to hear that it is boring (too much technical detail?) or not very clear (too much jargon, rationale not well articulated, research question not clearly linked to what you propose to do?). Don't be hurt; instead, thank your readers sincerely and get to work rewriting. Indeed, in a way the least gratitude is due to anyone who simply praises your application. (You feel good, but were you really helped in the long run?)

- Submit an Application that is Neat and Complies With Format Guidelines.

Remember that reviewers often have to read dozens of applications. Why make it difficult for them to read yours? Word process everything, and be neat. Don't use tiny type, compacted line spacing, or minuscule margins. Instead, edit yourself ruthlessly so you make the most important points without pushing the limits of the allotted space.

- Allow Plenty of Time.

Not only do you need to allow readers time to give your draft a careful review prior to submission, but also you need to plan on time for revision based on their comments. Indeed, if it doesn't take time to respond to your colleagues' comments, the draft probably hasn't had a thorough review, and you should try to find another reader (which also will add more lead time).

10.3.2 Research Grants

Nearly all of the advice for fellowships applies also to research grants; however, there are some additions and/or shifts in emphasis.

- Identify Several Possible Funding Sources.

In addition to information resources mentioned for fellowships, I would add your institution's office of grants and contracts (or sponsored research, etc.).

You might also place yourself on the mailing list or listservs for regularly published notices of grant opportunities. Many government agencies have such print or Web publications, and there is a growing availability of centralized websites for the federal government grant programs.

10.3.2.1 Request Current Program Guidelines

Experienced federal program managers note that some proposals do not fare well simply because they do not follow the suggested content guidelines. Often, subject matter guidelines for research grants are usually quite broad. Many program officers caution against artificially tailoring proposals to fit a particular year's priorities. Instead, make the strongest case you can for your area of research, clearly stating the hypothesis and explaining the broad context.

- Contact the Funding Agency.

In addition to the points made under fellowship applications, you should be aware that some agencies provide feedback on preliminary proposals. Indeed, some require such proposals. If agency policy allows such feedback, advice from program staff can be invaluable.

- Provide Both Context and Detail in the Research Plan.

The plan for a research grant will, of course, require more technical detail than for a fellowship application. Nonetheless, the research plan still needs to be presented in the context of a fundamental question, to have its importance explained, and to point out the gaps in knowledge that the project will address. The plan should also provide a thoughtful discussion of possible technical barriers and ways around them, and some preliminary data to indicate feasibility and promise. You must show that your research is a logical next step to previous work and that you have both the expertise and the materials the project requires.

- Prior to Submission, Seek Comment on Your Proposal.

Senior colleagues (with successful records of grant support) may be willing to critique your grant proposal. Seek two kinds of readers: one who is familiar with the research field and one who is not. The former will focus on detail and make sure the technical approach is reasonable. The latter will be able to comment on whether the writing is clear (free of jargon and of unarticulated leaps of logic), whether the hypothesis sounds reasonable, and whether a strong case has been made for proceeding along the lines proposed.

- If Not Funded, Ask for the Reviewer's Comments, Then Revise and Resubmit.

Many (but not all) funding agencies will give you a written summary of the reviewers' comments on your research proposal. Take the comments seriously – they may have identified an important flaw or gap in your plan. If you think they completely misunderstood the importance of the work and/or your research strategy, still accept much of the responsibility – you did not make your plan sufficiently clear.

Program staff who have heard all of the review discussion may be willing and able to convey additional information. Call them to discuss the review comments. They may amplify and suggest which comments should be taken most seriously. After due consideration of the feedback from review, revise your research proposal and submit it again, either to the same agency or to another one.

- Allow Plenty of Time.

Novices frequently misjudge the amount of time needed for others to read the draft and the amount of rewriting that will be required if the readers have been ruthlessly honest and critical (the kind of review you should make it clear you seek). Applicants should not submit a proposal until it is the best they can possibly make it.

If you haven't allowed enough time to revise, it might be better to submit a few months later if there are multiple review cycles in the year.

- Get Appointed to or Serve as an Ad Hoc Reviewer on a Study Section/Peer Review Panel.

The insight obtained by service on such panels is invaluable. Make it known that you would be available for such service. Many agencies publish panel rosters through which you can identify members you know. Don't be modest or shy: Ask colleagues to put your name forward. You can also contact the senior staff person for the panel and make your expertise and availability known (follow up a phone call with a copy of your c.v.).

- Be Confident and Be Persistent.

My final piece of advice to all fellowship and research grant applicants is not to be unduly discouraged or crushed by an apparent "failure." You may very well have been the applicant one slot away from an award in a very competitive program. Further, the peer review process is neither infallible nor exactly reproducible. Another set of reviewers may rate your application more highly. In addition, the revision, based on the review comments, will be more competitive. Remember that colleagues will readily talk about their funded applications but often forget to mention the one or two rejections that preceded the award. So be confident about your abilities and ideas and be persistent in your determination to find funding.

10.4 Keys to Success in Graduate School and Beyond

Virginia Walbot, Ph.D. in biology, has taught and enjoyed research in plant molecular biology and corn genetics at Stanford University, where she is currently Professor of Biological Sciences. Her laboratory utilizes a combination of techniques in molecular biology and plant physiology to understand maize. She is also active in scientific literacy efforts, and is developing new curriculum for non-science students and professional school students at Stanford.

10.4.1 Choosing an Advisor

The foremost criterion for choosing an advisor is shared interest in a scientific problem. What specific subjects in your field interest you the most? Which technical approaches are most intriguing? When you have defined your intrinsic interests, talk to potential advisors, postdoctoral mentors, and colleagues about science. Use every opportunity to do so throughout your career. For example, in graduate school you will take courses and, probably, either do lab rotations or enroll in graduate laboratory courses. Use professorial office hours to ask questions about extra reading, to discuss potential research projects, and to get answers to your specific questions. If you are a rotating graduate student, set up a directed reading program with your professor and meet privately with him/her every week or so to discuss the scientific background and key papers relevant to the lab's work. If you are already a professional, make appointments with key people in your field and keep in regular touch via telephone, by electronic mail, and/or through personal visits.

Direct conversation with a lab director about subjects of mutual scientific interest will allow you to best judge his/her style and how fruitful your interactions with this person are or could be. Are your ideas encouraged or discouraged? Is the competition respected or put down? Does the professor quote with pride the individual contributions of people in his/her lab or merely state that "my lab did..." You should gain insight into this individual's expectations about work habits and schedule, two important things to know about before joining a lab. And, look carefully at the fate of "graduates" of the lab: Do you want to end up with their kind of postdoctoral appointments or jobs? What percentage of students finish PhDs in this lab, and how long does it take? Are men and women students treated equally – given credit for their ideas and encouraged to take on equally challenging projects? Do the lab members receive support from this advisor and from each other?

10.4.2 Joining a Lab

Participate actively in group meetings and other forums for intellectual/theoretical discussions that provide direction to your lab. It is not sufficient just to do a good job at the bench. After all, technicians can master technical skills. Good scientists are recognized for their abilities to frame questions, to pose potential solutions, and to inspire others to pursue these solutions. Practice these skills through lively discussions, defend your ideas, and help positions evolve. Think about best and worst case scenarios creatively. Don't be afraid to sound overly ambitious!

Arrange a weekly or biweekly private meeting with your advisor, even for 15 minutes, to discuss what you are doing (or not doing). Without being offensive, take charge of the meetings and make a presentation on an experimental strategy, a key paper you've read, or the data you've generated since the last meeting.

Prepare for each meeting as if it were a job interview. Ask yourself the tough questions before you reach the professor's office so that when/if they come up, you will be able to give thoughtful answers. Know the relevant literature and bring those reprints with you. And, if the professor doesn't recognize a potentially fatal flaw, demonstrate that you are self-critical by raising it yourself. Similarly, think of one or two additional experiments that you could do or should do, based on what you will be presenting. Over time, professors expect students to become expert in judging what's hot and what's not and to plan more and more of their own work. A professor forced to suggest key controls or to point out an obvious lead to an advanced student will not be impressed with your intellectual development.

Over time, the responsibility for your project should shift from a supervised to a collegial relationship, with you taking the lead in experimental design, analysis, and planning. Accept this change and thrive on the trust that the faculty shows in your abilities. Schedule a few hours per week to think about your research problem and to make a list of as many projects as you can. Then, cross out the merely good ideas, think critically about the remaining very good ones, and winnow these to one or two excellent ones. Next, spend some time in the library and with lab friends discussing these potential new directions and the types of new techniques that you would need to implement or invent.

You must invest in your own intellectual progress by thinking as well as doing. It is common for students to become comfortable with a project (often suggested by a faculty member) but, after an initial flurry of reading (as if preparing for a term paper), to cease to think critically about the project. This is a serious mistake. The project should be evolving as you work on it – you should analyze every piece of data for its obvious message and its various what if scenarios. Read widely and think hard. The initial conception of the project may and probably should change, as you assume more of the responsibility for doing as well as for thinking about the project. In the end, the best students have worked on an original problem, having taken an initial (perhaps even ill-conceived) faculty suggestion, and have nurtured it into a contribution.

As you meet routinely with your advisor and impress him/her with your intellectual commitment and abilities, your level of comfort in bringing up bad news should increase. Your advisor formed a good impression of you before agreeing to let you work in the lab, and your consistent good preparation for meetings should have solidified this impression. Now, when you have technical difficulties or think you should change projects, you should have a sympathetic listener.

10.4.3 Meeting With Your Thesis Committee

This committee will not only advise you during graduate school but individual members will also write letters for fellowship applications, for postdoctoral recommendations, for your first job, and possibly for your tenure or promotion. Choose members with the clout, the time, and the interest to help you now and in the future. Hold meetings with your committee often – more than the minimum, but not so frequently that you pester busy and important people. Each meeting offers an additional opportunity to impress a captive audience. As with your meetings with your advisor, rehearse what you are going to present before each convening of your committee. Practice your presentation in advance in front of an audience (fellow students, your labmates).

Give a polished performance, and be sure to let committee members have time to ask you questions and make comments. Distribute a written outline or agenda, and make sure to thank committee members for their time. As your results come in, add people to the committee who reflect your evolving interests. Keep all members informed of good news such as manuscripts accepted for publication and your presentations at meetings. Drop by to chat with committee members or send them a periodic e-mail that starts "surprise, for once I don't have a request, I just wanted to let you know..." Also, remember that everyone loves to hear that you noticed a recent publication or honor. Faculty are very flattered when students or postdocs from other labs comment on their work.

At the postdoctoral stage and beyond, budget your time so that you can volunteer for or accept invitations to write short reviews. When you read a particularly exciting paper or one that provides methods of potential use to your lab group, make an extra copy and give it to your advisor or post it in the lab. Then, make an announcement about the paper at the lab group meeting. Carry out lab duties as a team member. There are few tasks that women can't do; however, don't be a wimp about asking other people to clean up and do their share. It also never hurts to bake cookies for a hungry lab or to arrange a kayak trip or other activity that you are comfortable doing.

10.4.4 Some Final Tips

You cannot expect objective, instant recognition of your success – there are no grades for everyday life. Mary women students do obtain excellent grades throughout undergraduate and graduate school, but the importance of these grades declines with each additional day you are a professional.

Recognize that true success in science rests on your abilities in your field, not on your preparation to enter it. And, in many cases, recognition of your contributions can take a long time – months, even years, can pass between your key discovery and the world's catching up to recognize it.

Don't undervalue your contributions. Women often give themselves a barely passing C in comparison to the stars in their field, while men with the same credentials view themselves as As, based on their potential for success. Be realistic in your self-evaluation. Then raise your grade because you are committed to make the extra effort to be a success.

Remind yourself that you love science and that the challenges are opportunities. And remember that every person has unique contributions to make. Great ideas – the original and important changes in a field – very often result from the insight of a single person.

10.5 Building Confidence and Connection

Sheella Mierson, Ph.D. in biophysics, is President of Creative Learning Solutions, Inc. Her company offers consulting and customized training to organizations to help them develop effective people skills for the workplace, with a focus on multicultural issues, leadership and team development, and group problem solving. Francie Chew, Ph.D. in biology, is Professor of Biology and Director of the American Studies Program at Tufts University. Her research interests focus on ecological consequences of encounters between native insects and naturalized plants.

> What kind of work do you do?
> I'm a scientist.
> You must be smart! I was never very good at science in school.

Sound familiar? As scientists, we spend our lives thinking in areas where many females are taught they shouldn't and/or couldn't venture. When women in other lines of work meet us, out come their doubts about their own intelligence. Meet internalized sexism.

Working toward gender equality in science requires efforts on two fronts. The first, activism to change external conditions, is the subject of other chapters in this volume. Learning about examples of gender-equitable policies and successful intervention strategies helps us articulate a vision and draw inspiration from the experiences of others.

Second, in addition to changing institutional policies and practices, work to change attitudes is necessary; otherwise sexist institutions and policies can return and become perpetuated in other guises. This chapter focuses on this second area, in particular how female scientists' confidence and connection with others contribute both to improving our personal responses to external events and also to changing external factors. We view efforts in these two areas – external conditions and attitudes – as complementary and essential to establishing gender equality in science and elsewhere.

10.5.1 Sexism, Internalized Sexism, and Stereotypes about Scientists

Sexism is an institutionalized system where females and males are treated differently on the basis of gender. There is differential access to resources and to the power to make decisions about these resources. One gender is targeted for exclusion or exploitation, while the other gender implements or enforces this systematic mistreatment. None of us chooses our role in this system, but we are all socialized as we grow up to conform to gender roles. When this socialization includes misinformation about our abilities in science and math, we can acquire distorted, gender-specific expectations. These expectations influence how we treat ourselves, each other, and members of the other gender in relation to science and math. These attitudes can result in internalized sexism in women (as members of the group targeted for exclusion or exploitation) and sexist attitudes and behavior in men (as the group assigned to implement the system).

An important aspect of internalized sexism for women scientists is that many of us have nagging doubts about our intelligence and competence, doubts which may persist despite all evidence to the contrary. Sheila Widnall (Widnall 1988) documents studies showing that female scientists' self-esteem drops relative to that of males in high school, undergraduate, and graduate school. This occurs even though males' and females' academic records are comparable. Studies summarized by Gloria Steinem (Steinem 1992) indicate that this drop in self-esteem occurs for females relative to males in all fields. In high school and college, even when women's records are better than men's, men's expectations for themselves are higher (Ruskai 1991).

One reason females may learn not to take themselves seriously is that they are often not taken seriously by teachers (Steinem 1992). Numerous studies show that boys are called on more often and talk more in their average response than girls, yet in one study when teachers were shown films of classroom discussion in which boys outtalked girls by a ratio of three to one, the teachers – including feminists – still perceived the girls as talking much more. We are so culturally trained to think that females talk too much, that we should be good "listeners," that we seem to measure ourselves against those expectations, not reality (Steinem 1992).

Such expectations are pervasive even among well-intentioned people (Dovidio and Gaertner 1998), apply to females in science at all levels (Sonnert and Holten 1996; Fox 1999), and are a major reason for slow progress toward gender-equal institutions (Valian 1998).

An interesting complication occurs when stereotypes about scientists intersect with internalized sexism. One such stereotype is that scientists are more intelligent than other people. This belief, which has no basis in fact, keeps the latter convinced they are other than smart and powerful. A second stereotype is that scientists have fewer needs or skills for close contact and nurturing than do other human beings. This second piece of misinformation seems to affect men and women scientists in different ways. For men, it reinforces the conditioning that they are supposed to be emotionally independent and that females can think better than they can about relationships. For women, something different happens. In our society women are regarded as the nurturing sex. Indeed, some consider that the responsibility for nurturing is ours alone, and that only we can be successful at it. If women scientists are seen as being less nurturing and having fewer needs for close relationships, however, we are set apart from other women. This distance can contribute to whatever isolation we experience as female scientists.

Internalized sexism and the cumulative effects of unequal treatment have some particular effects on us as women scientists. In addition to compromising our self-confidence and our ability to remember our accomplishments, these attitudes also impair our ability to see connections and shared commonalities with others, including other women scientists. Being convinced we are isolated, in combination with any internalized doubts about our competence, can make us tend to take difficulties personally.

Any sense of isolation stems both from what we carry inside as internalized sexism and from external isolation resulting from our often low representation and unequal status in science. Being connected to each other makes it easier to remember that many of the difficulties each of us faces arise from something other than personal defects. Differential treatment from multiple sources (from intersecting identities as members of other groups that receive unequal treatment in our society, such as working class, women of color, lesbians, or older women) can further reinforce isolation and undermine self-esteem. When we realize what we have been up against, we can give ourselves full credit for persevering and flourishing as women scientists.

The attitudes of sexism and internalized sexism reinforce and perpetuate scientific institutions and policies that disadvantage females relative to males in representation, status, and treatment. Changing institutional policies and practices is half the work; also doing the internal work makes us less vulnerable, better able to respond effectively to any unequal treatment we encounter, and better able to connect to each other and to allies.

10.5.2 *Claiming our Intelligence, Confidence and Femaleness*

The reality is that both men and women flourish when they feel good about themselves and about their intelligence, and when they are surrounded by the close relationships they desire. Each of us is wonderful – a smart, unique person, woman, and scientist. We can keep the human ability to have loving, nurturing relationships and still enjoy another part of being human, namely curiosity and intelligence. The reality about women is that we can be fully contributing scientists and also be fully female.

We recommend four approaches to claiming our voices and assisting others in doing the same:

- As part of a support network, set up a regular meeting with a friend.
  Get together once or twice each week; a half hour to an hour in person or by telephone is a good start. Agree to socialize some other time, and to keep confidences. Each person takes a turn (10 to 25 minutes; we use a laboratory

timer). When it's your turn to talk, focus on your achievements for that week, talk about what's been good and what's been challenging, and think out loud about next steps for yourself in any area of your life. It is fine to talk about your feelings, laugh, cry, yawn, rage, sweat, and tremble. These are physical manifestations of emotional release; you will then be less overwhelmed by feelings and better able to think clearly.

When it's your turn to listen, do so attentively and supportively, and without interrupting with questions, comments, or advice. Give each person a chance to set her own agenda. At the end of each person's turn to talk, make sure her attention refocuses on some benign aspect of the present, such as counting how many items in the room are a certain color. One piece of advice: If you drink alcohol, do that at a time other than when you do these dyads. Physiological changes occur when you release feelings, resulting in clearer thinking. When alcohol is present in the body, these mechanisms function differently and the benefits are less. In some cultures people have permission to show their feelings primarily when drinking, so this means giving ourselves permission to do so at other times.

This dyad format will help put you in charge of your decisions to choose, seek advice, and act. Other uses of the dyad format will be suggested below.

- Remember the times when you (or other women) actively resisted or defied sexism and internalized sexism, either by standing up for yourself or by assisting others.

  This recounting of personal (and collective) history will remind you of your courage, passion, caring, generosity, and resourcefulness. It will remind you that you have been active, that you have chosen to stay and fight or negotiate, that you have persisted. This remembering sometimes elicits feelings, whose expression and release is to be encouraged as part of the emotional healing process. You may choose to use your turn in a dyad session on this topic.

- If you teach, promote active participation by all students.

  Engage "shy" students with eye contact. Request questions or responses from people who haven't spoken in class recently. Structure some thinking activities using student dyads to encourage reflection. In these "think-and-listen" activities, one student takes a turn for one or two minutes while the other listens; then they exchange roles. This approach is similar to the longer dyad described above, but usually the assignment focuses on a specific topic such as "What did I just hear that is new to me?" after which you can ask for questions.

  Expect the best from your students, even if their test scores in a large introductory course were lower than they would like. Expect that the females as well as the males will engage in challenging work, compete for scholarships or internships, and have high aspirations. A teacher's self-awareness is helpful in applying these approaches. Tufts biology professor Saul Slapikoff describes his discovery, through observations made by his colleague Sara Freedman, of his own unconscious focus on men, despite the fact that his

course included a major section on debunking scientific myths about sexism and that a slight *majority* of the students were women (Slapikoff 1985).
- Support your colleagues by noticing when they do things well.

Tell your colleagues when you learn from their seminar presentation or from their questions. Tell them when students compliment their teaching. Nearly everyone, male and female alike, has been consistently told, "You haven't done enough" and "You could do more or better." What a boost to them when someone notices they've done something well! You will probably find, in addition, that others tend to notice and appreciate your work in return. Further, you may be consulted more often and more openly, resulting in better connections for yourself.

10.5.3 Building Our Connections to Others

Historically, women who have successfully challenged oppressive institutions and policies have usually been aided by allies – men and a few women in high places. Because allies can often exert powerful influence to change external conditions, our goal is to persuade every person in a leadership role to become an ally. Because potential allies have been socialized to gender roles and leadership roles in institutions, it may be tempting to see them as enemies to progress. We observe, however, that mostly they are good people doing the best they can given the pressures the system puts on them, who could sometimes use a more inclusive picture of scientists and science, and who could benefit from a larger picture of their own significance.

We recommend three ways to make connections with potential allies and help them to include women in their thinking and to recognize their own significance.

- Sincerely appreciate what they have done well; it gets their attention.

Your engagement with them as human beings counteracts stereotypes about women scientists. Use your support system and dyad sessions to remember that these potential allies are good people who could stand to be more aware of their own power, who may retreat to "obeying the rules" when challenged. Use your dyad sessions to remember the times when allies acted on your behalf, or when you acted as allies to someone else, and to express and release any feelings that arise. This will help you remember more easily that you actually want to cultivate a connection with someone who seems different and more powerful than you are.
- Engage potential allies in discussion of the issues by both listening to their thoughts and challenging them to consider alternative information and ideas that reflect gender equality or more imaginative ways to implement policy.

Tell them that a gender-equal system includes females in its design from the outset, and goes beyond females trying to adapt to a system originally formulated without reference to their participation. Challenge the system without blaming them, positioning yourself as loyal opposition rather than as an adversary. While this activism takes thought and energy, it is effective in staying hopeful and maintaining a sense of being powerful. Use your support network and dyad sessions to process feelings that come up when you assert yourself, to review what went well in these interactions, what you might have done differently, and what your next steps might be.

- When you are with allies in high places, women or men, remind them that they are good, caring human beings.

Encourage them to remember times when they resisted pressure to exclude others (for example, resisting a dare to taunt another child at school). Encourage them to remember the times that they and people like them took risks to act on behalf of others, and times that other people came to support them. Remind them that human beings make mistakes and can often correct them. These memories sometimes elicit strong feelings, whose expression and release you can encourage via laughing, crying, yawning, raging, sweating, and trembling.

10.5.4 Towards Gender Equality

All of us, as we rise in the system, are subject to the isolation and pressures felt by people in high positions. If we aspire to leadership we will be similarly vulnerable unless we actively work to stay connected to others and to claim our own intelligence and voice. For this reason, continued work on the feelings – in the context of a support network – is a good way to prevent the burnout that might otherwise occur as we become more visible. We recommend processing feelings as one way to change attitudes; otherwise sexist policies can re-enacted in other guises.

Science developed within societies that were and are sexist, and a large body of literature documents the persistent inequality of representation and status of women in science. As female scientists, we have much to contribute to science – if we get the opportunity to do so. Our individual initiatives will strengthen our own connection and self-confidence. Our collective initiatives also contribute to more pervasive attitude changes in STEM fields, so that inclusion becomes part of the culture of scientists. Everyone wins from such inclusion. Improving the professional and human climate of our scientific institutions benefits all concerned – both women and men.

10.6 Helping Those Who Follow

Betty P. Preece, B.S. in electrical engineering and M.S. in science education, has been involved for many years in career guidance for females of all ages through the Association for Women in Science, , the Society for Women Engineers, the

American Association of Physics Teachers, and other organizations. She works as a teacher and an engineer.

Who is guiding our girls? Who is giving them ideas about what they could do with their lives? Who is helping them make plans and set goals for their futures? The answer seems to be "everyone" and "no one."

Many who come in contact with girls send them some kind of message about their lives. It may be accepted as appropriate, rejected as undesirable, or absorbed for later unconscious recall. The message may be spoken or merely an opportunity offered to observe more thoughtfully. It could say, "You could be doing what I am doing!" or it could ask, "Do you really want to be as I am?"

Much too often, girls have no one actively exposing them to careers that produce a vision of and a path toward a satisfying future. Only some girls realize that nearly all of them will work for most of their adult lives. Their work can be drudgery[,] or it can offer rewards in self-esteem as well as money.

Readers of this handbook are among those who can provide the kind of career guidance that will make the difference for girls. Although we cannot reach many girls individually, we can help bring awareness directly to some of them and indirectly to others by reaching those who do influence them – their families, friends, teachers, community members, and the media, for example.

10.6.1 How Are Girls Being Guided?

A great many things affect girls' career plans or lack of them. Sometimes the influence is planned, but most often it is not and goes unrecognized as a factor in forming lifetime goals. Girls consistently rate peer pressure as having significant effects on their career choices of school classes and extracurricular activities. Educators and counselors present career images through lessons, texts, comments, resource materials, and role models. Recreation leads many students to try careers that emulate those of popular figures in sports, music, and other performing arts. Families, friends, and the community all serve daily as examples of the work force.

Television programs are powerful sources of good and bad career information. More and more students are being touched by specific programs designed to provide enrichment in the sciences, engineering, or math. Still, too many are not receiving information that enables them to make their own viable choices.

We women in STEM-related careers want to see girls – and boys, too – become aware that education can be the key to a better life through planned careers dependent upon choice rather than chance. Only with wider exposure to career information, better understanding of their own capabilities, and knowledge of opportunities open to them can the young make this choice meaningfully.

10.6.2 Opening Doors for Girls

We can widen girls' choices by making it easier for them to take part in the activities in many arenas that influence career choices. For instance, we can encourage

- networking
- mentoring and role modeling
- curriculum changes
- enlightened counseling
- community awareness
- abolishing bias against science by parents, youth, and the public
- better representation in both the popular and the scientific media
- family-based science activities
- increased and widespread contact with science and technology workplaces
- support for postsecondary and vocational education
- special programs targeted to give girls experiences that will expand their career outlooks

Spring and early summer are the best times to plan the activities you and/or your group may be able to implement in the coming school year. To raise girls' consciousness about their options, organizations like the Association for Women in Science, the Society of Women Engineers, the American Association of University Women, or technical societies have active career guidance programs and resource materials for girls and women. Make a list of programs that you think were successful last year, and gather information on the ones you think might work in the future. Then, stack them against your resources to choose the ones you and/or your organization can best support. After your plans have formed, line up help from your chosen group (for example, your Association for Women in Science chapter) before fall schedules fill.

Organizations that can help include school district science coordinators (many of whom work all year), science centers, Girl Scout Councils or girls' clubs, professional organizations, women's groups, and churches with youth programs – especially minority churches, which often have strong community outreach programs. Tell your chosen group what you would like to see happen, how you can help, and see how your priorities relate to theirs. Then jointly proceed to make and carry out plans.

Plan ahead for the regional science fairs that will be taking place in February and March and for the state fairs soon after. If you want to help students, judge, or provide prizes, contact your school district's science coordinator.

Even if you are not part of any formal organization, contribute as much as your resources permit.

10.6.3 Thinking Globally

I have talked with educators, scientists, and engineers from many countries at conferences on three continents. We discussed kindergarten through graduate school science and math curriculums, activities to attract students into these courses, and strategies to increase the low participation of women and under-represented and under-prepared groups. The issues and the attempts to find solutions seem similar worldwide.

Women in many countries are eager to exchange research and information, especially with their U.S. counterparts. Though I disagree with them, many women in science from other countries don't believe they have much to offer American counterparts, whose significant progress they much admire. The discovery of the common ground among women in science and science education throughout the world provides a strong bond for all, not just for those women who are fortunate enough to attend the international conferences that contribute invaluable opportunities for networks.

I attended a workshop for counselors at the school district and college level who were working with sex equity programs for single parents, displaced homemakers, and unmarried pregnant women. The workshop was formed to help the counselors to increase awareness of more and better career opportunities for such women. Preparing for such careers by taking math and science classes to give wider options was frequently stressed in the sessions. During a discussion, one of the counselors remarked that she was giving women wider options is by advising them to take good liberal arts courses. Others tried to show her that girls who don't take math and science are closing the doors to many career paths. We could not convince her that the liberal arts alone do not provide a background that will lead to meaningfully broad career opportunities. We still have a lot of work to do with those who are advising young women about career preparation.

In her address to the Ninth International Conference of Women Scientists and Engineers in England in July 1991, Baroness Platt of Writle embroidered a new version of the old saying, "Diamonds are a girl's best friend." She continued, "But they can be sold only once. Science skills can be sold every day for all of one's life."

10.6.4 How Well Are We Guiding?

Unfortunately, evaluating efforts to provide girls with career guidance is difficult. Most of the results occur over a lifetime, and with so many variables it is impossible to determine accurately the influence of one in isolation. Some measurable outcomes indicating changes for the better include increased enrollment of females in science and math courses at the precollege, college, and postgraduate levels and greater women's employment in science, engineering, and math jobs. Most of these indicators record show increases followed by periods of little growth. It is difficult to assess whether attitudes of parents, peers, and the community have changed to provide more support for girls pursuing nontraditional careers. These uncertainties do not mean that we should abandon efforts to guide girls. Indeed we must continue in spite of the fact that we cannot always measure the significance of our attempts.

In looking over the past decade, I wonder if the actual process of mentoring has changed? It appears that not only has the process changed but also the delivery. In the next few paragraphs we will look at some of these changes which are definitely affecting mentoring.

The ages of the girls being contacted has shifted downward to the elementary school years. If our awareness interventions wait until senior or even junior high school age, the patterns and educational preparation for career choices have already been set – frequently in directions other than science. In addition, girls and women of all ages seem to continue to be vulnerable to real or perceived peer pressures that significantly affect career choices.

Engineering college enrollments are decreasing but the percentage of women has been staying relatively unchanged. Hence, many engineering colleges in the United States seem to be putting their efforts toward the same "balanced" enrollment.

Many recruiting and retaining programs today seem to be struggling to provide activities that rival television and multimedia productions. Simple activities which are explained so that girls really understand the science principles behind them are more likely to have a lasting effect. Then students can take them home to replicate for their families and friends. When women lead these activities, there is double value. This kind of science helps everyone know more about career options and preparation.

The timeline for planning and funding has moved earlier in the year. Spring and late summer are too late. Your planning should begin 9 or 10 months before the date of the activities so that you can secure facilities, pick prime dates, and seek funds while next-year budgets are being prepared.

Mentoring – especially long-term face-to-face mentoring – continues to prove its values in encouraging young girls and retaining women in both college and the workforce. While the mentoring during special activities provides significant awareness, the sustained I-care-about-you contacts of long-term role-model mentoring builds successful rapport. Electronic mentoring has become very popular, particularly since it can occur at times and locations easily accessed by both mentor and protégé. Women who work in environments where there are few other women scientists or engineers often find such mentoring very valuable. A number of national programs provide this but many local professional organizations and colleges are equally effective with email mentoring of girls and women.

Perhaps the most effective mentoring occurs during internships and especially during co-op college programs. Here the sustained daily contact in a wide variety of situations offers many kinds of opportunities to interact with not only a mentor but also workers in many kinds of careers. Young women scientists increasingly highlight their concerns for balancing work and family, for avoiding isolation, for enhancing opportunities for mentoring.

Thus, considering these changes, what seems to continue to be the most valuable techniques for mentoring? Long term face-to-face mentoring by women role models – but remember, any contact you are able to make can lead to a life changed forever. You probably will never know the impact you make; however, if you never make any contacts, you'll never be able even to think you made a difference. Be a mentor and make a difference!

10.7 Professional Responsibility

Stephanie J. Bird, Ph.D., is a past president of the Association for Women in Science, under whose leadership the organization began development of its strong mentoring programs. Trained as a neuroscientist, her present interests emphasize the ethical, legal, and social policy implications of scientific research in general. She is a private consultant to educational institutions on the responsible conduct of research and the professional responsibilities of scientists.

As science professionals, we have responsibilities in addition to those we have as adults, parents, and citizens. These additional responsibilities come in various shapes and sizes but fall into two general categories: Those that preserve the integrity of the profession and those that stem from the role that science plays as a component of society as a whole.

As researchers, we must be concerned about the accuracy and reliability of our work. Science is built on the work of others. We count on the veracity of what has been done before. If it is incorrect, not accurately reported, or misinterpreted, then the time and effort, supplies, equipment, and animals that the research has required were wasted.

Intentional fraud, however rare, tears the fabric of science. Yet even sloppy science is unethical. If I carelessly contaminate stock solutions or jars of chemicals and drugs, if I don't clean up shared work space, if I fail to return tools and equipment to their designated storage locations, then I waste the time and energy of others. Worse, I expose my research and that of co-workers to unnecessary errors, and potentially all of us may face additional health risks.

Collaboration requires trust. We depend upon colleagues in fields outside our areas of expertise to take the same care as we do to insure that joint efforts are worth of the energy and resources invested in them. But trust is not enough. We need to take responsibility for the reliability and quality of the work presented in papers that have our names on them. The Council of Biology Editors, as do many other professional groups, requires that each author be able to explain the rationale for specific research procedures, that is, be able to say why and how the research was conducted, and why the conclusions follow from the results. This responsibility is the natural counterpart to taking credit for the work that bears our name and follows reasonably from participation in the development and design of the study, in the collection and analysis of data, and in writing the paper. This high standard does not acknowledge (let alone accept) the all too common emphasis on quantity at the expense of quality, resulting in the phenomenon called the least publishable unit (Broad 1981). Yet, to accept lower standards wounds our profession and ignores the issue of the interdependence of research. Low standards reflect negatively on all of us. Insistence on high standards is an extension of our mutual and shared responsibility to maintain the quality and stature of our field.

10.7.1 Literacy and Expertise

In our role as educators, we have a two-fold responsibility. The increasing importance of science and technology requires a citizenry that can grasp fundamental scientific concepts that may run counter to intuition or popular wisdom but are essential to making intelligent decisions and choices in a complex and dynamic world. All too often, however, concepts in math and science are made unnecessarily esoteric, as if to serve as a filter to create an intellectual elite who can, and thus should, make policy decisions that involve science. As educators, we have a responsibility to bridge gaps in experience, background, and socialization to make clear key concepts in science for as many students as possible.

At the same time, we have a responsibility, to thoroughly educate and train students pursuing careers in science in the principles and techniques of the field. We need to set high standards for them and for ourselves as role models. But our responsibility to these students extends beyond teaching them the rudiments of science: It includes educating them about the skills, conventions, and mores of the field, as well as their responsibilities as science professionals.

10.7.2 Remembering the Big Picture

Future science professionals need to be reminded of the implications of the basic methodology of science. We need to step back and see that the categories into which we divide our world, and our science, are useful but limited and limiting. We, simplify to understand and order our world, to make it possible to function, cope, and progress. In science, we order the conditions of an experiment, eliminate as many variables; as possible, in order to solve an equation and assess the influence of various factors on an outcome or process. But our method is artificial and an artifice. It is one of the myriad ways we modify reality and "vex Nature" in order to make it reveal its truths.

It is important to keep in mind that this simplification is a means to an end and is only meant to be an interim step. The system under study needs to be restored to, or reexamined in, its unmanipulated state so that the function of other factors can be investigated. Unfortunately, too often this crucial step is neglected or intellectually postponed, and we lose sight of the fact that an artificial situation has been created. Yet we have a responsibility to ourselves, our colleagues, science, society, and our students to recognize and address the fact that this is one of the ways in which science is done, that is, we set tip artificial situations and perturb the system to see how it works. The ways in which we change things, how we select and delimit variables to be tested and examined, reflect assumptions that are often neither acknowledged nor obvious. But they need to be identified and defined. It is critical that we take serious notice of the wider implications and potentially unintended consequences of this process.

A case in point from the life sciences is the general use of male animals (and exclusion of females) in basic research except when primarily or exclusively female physiology or behavior is being investigated. While this convention in research design was adopted to eliminate the variables of cycling female hormones, experiments have rarely been repeated in females to determine the role of female hormones in the phenomenon and the applicability of research findings for females. This practice reflects and propagates the underlying assumption that only female hormones cycle, perpetuates the erroneous as well as insidious notion that females are simply modified males, and unnecessarily and unethically puts at risk females involved in clinical trials and/or female consumers of compounds, devices, and procedures that are the fruits of biomedical research. This weakness in experimental design, which leads to a serious gap in scientific knowledge, has not been fully addressed. This lacuna is evidenced by women's invisibility in textbooks, funding priorities, reviewers' comments, and student training.

10.7.3 Helping Women Entering Science

As women science professionals, we have a special opportunity to address the particular obstacles and concerns of women exploring and pursuing career options in the sciences. Barbara Furin Sloat highlights a number of ways we as science educators can have a positive influence on young women and "break the feedback loop of lowered expectations and lowered self-image that leads to discouragement, loss of motivation, and lowering career ambitions." In addition, we need to help women entering science see a clear picture of the realities of their future careers so that they can determine how best to proceed in their academic choices. Providing such an overview is not easy, because each of us has individual perspectives and experiences that differ widely but that inform our view of reality. Yet, by sharing those perspectives and experiences, we can help women entering science better understand the challenges, obstacles, and rewards that lie ahead, even though both students and professions are changing with a changing society.

As women science professionals, we bring our own experience to these issues. But one need not be established in one's field to serve as a mentor. Mentoring is, in part, sharing our experience and our expertise; it is identifying what we have learned and sharing it with others. Upper-level students can share their experience of courses and instructors with lower-level students; graduate students know what they have learned about selecting graduate schools, thesis advisors, and thesis topics; postdoctoral fellows know about these topics too and about the process and pitfalls of selecting a position.

We are all in this together. We have much to gain, as well as a responsibility to those who have smoothed the way before us.

10.7.4 Working With Our Colleagues

Women science professionals also have a responsibility to each other. In large measure, organizations such as the Association for Women in Science and the Society of Women Engineers are expressions of the recognition of this responsibility. These organizations work to articulate the concerns and goals of women in STEM. We join together – combining our talents, skills, and knowledge – to address central issues. At the national and the local levels, in our institutions and local chapters, each of us must do what we can, voting, serving on committees, and taking leadership roles whenever possible, to assure a positive, forward direction. We cannot leave these tasks to everyone else.

10.7.5 Serving Society

As science professionals, we have a responsibility to see that science is not misused, that it does not become a tool for abuse. This is not easy, and it can take time. Often, as scientists, we feel that social commitment is "riot our job," and that we are not good at raising and addressing concerns regarding the use of science in public policy. But in our own fields we are far more likely to know of the limits and conditions of our knowledge and, more to the point, the problematic aspects of using this information as the basis for public policy. In addition we recognize – or are most able to sort out and identify – underlying assumptions. By identifying and acknowledging the role of potentially value-laden assumptions in science, we emphasize the way in which social values enter science. We have a right and a responsibility to make cur voices heard when our science – our contribution to society – is being misquoted, misunderstood, or misapplied.

As science professionals and as citizens, we have a right and a responsibility to speak out about the effects of war and other destructive applications of technology. Whether the impact is on the environment, irreplaceable archeological sites, the psychological well-being and development of generations, or funding of education, research, and health care, we must speak out on issues; and reality as we see them.

Chapter 11
Provocative Thoughts for a Better Future

> *Those who cannot remember the past are condemned to repeat it.*
>
> George Santanaya

11.1 The 'Problem' of Women in Science: Why is it So Difficult to Convince People that There is One?

Sheila Tobias, an early member of the new feminist movement and a political activist, was a pioneer in women's studies and in the identification of math anxiety and science avoidance as important and surmountable barriers for bright women. She has written numerous books on these subjects and others, taught and done administration at the college level.

I have spent the past 20 years as a feminist analyst and activist – in that order – because analysis must precede activism if policy recommendations are to be effective. Most recently, I have been studying the ideology of science and science achievement. That study has led me to link the ideology of patriarchy to the mythologies that dominate the search for new scientific talent.

11.1.1 Patriarchy

From a feminist point of view, patriarchy is the domination of a society by males whose primary purpose is to construct and to maintain a certain power over women. In such a society (and we feminists believe that we live in one), much of what we take to be traditional or even "true" is really an extended political maneuver to maintain the unequal power relationship between males and females. This does not necessarily imply a conspiracy, in which every man colludes with every other man to keep women out of power. Rather, in their enjoyment of privilege and individual advantage, most men accede to the system that is in place and have no particular interest in making change.

This theory was first sketched out in modern times by theoretician, who observed that there were three dimensions on which males and females are differentiated: temperament, role, and status (Millett 1970). The man on the street and the traditional psychologist will both typically assert that women are temperamentally, that is, psychologically, different from men. Women, they believe, are more passive and men are more active; women are more dependent, and men, more independent. If such theories are true, this means that women are less comfortable with and less likely to seek a life of their own. In terms of rationality, these theories contend, women are intuitive and rely on their feelings for truth, while men are more concerned with what is demonstrably true and are more systematic in their thinking. Consider what such "sex differences" mean for women in science.

As to the second dimension – adult roles – patriarchs want people to believe that adult role differentiation is as "natural" as (and, indeed, grows out of) temperamental differences. The mothering desire to marry and to make a home is as natural for females as the males need to make his mark in the outside world. Woman is a private entity; man a public one.

Finally, as an equally inevitable consequence of role differentiation, men enjoy higher status than women, because what men like and do is socially more important than what women like and do.

Virginia Woolf once observed that Leo Tolstoy was considered a greater writer than Jane Austen because he wrote about war and peace, while Austen limited herself to dramas of interpersonal relationships. But who decides, Woolf asked, that war is more important than interpersonal relations? Someone else – the society in which the novelists wrote and are read. Since men derive status from what they do, the argument is cyclic: What men do is more important because men do it.

In the popular view, to sum up, the causal chain of temperament-role-status begins with temperament and ends with status. Temperamental differences, which supposedly coincide with birth, lead inevitably, logically, and naturally to role and status differentiations. Since the key years of professional advancement (in our era) are between 25 and 40, the years when "normal" women are fulfilling their mother-home yearnings, men end up in higher status positions than women.

11.1.2 Upending Traditions

Thus, Millett explained how patriarchy works. But her more important contribution was simply to reverse that causal chain, much in the way that we tell our political theory students that Marx stood Hegel on his head. Status, argued Millett, comes first. For patriarchy, the essential task is to maintain the superiority of males over females. And to do so, to keep women out of power, women are assigned roles that busy them in caretaking in isolation from other adults.

How are these different roles assigned? And what causes women to accept them? With brilliant insight, Millett defined the people-professionals (social and behavioral scientists, therapists, and educators) as being a new "priesthood" exercising social control by redefining what is "normal" and what is not. Let a woman declare the mother role constricting, and she is found unnatural, abnormal, and in need of professional help. And that, Millett helped us see, is the modern-day punishment for inappropriate role behavior.

Because it sheds new light on so many previously accepted traditions, and because it made us angry, Millett's analysis was a powerful motivator for the new feminism of the 1970s. Her assertion had an impact like that of the Copernican revolution. It evicted from centrality conventional assumptions about women's temperament, roles, and status, as Copernicus did the earth when he set the sun at the center of the solar system. Millett influenced many women in the early 1970s to look at the arrangements they had made with life, their relationship to men and other women, and the result was often the "click!" experience: When women suddenly saw their circumstances in these political terms, something went "click!" and life was never the same thereafter.

11.1.3 Implications for Women in Science

I think that the temperament required for dedicated, original work – the staying-up-all-night kind of work, the I-can't-think-about-anything-else-darling-not-even-you-tonight, because I've-got-those-things-growing-in-the-petri-dish kind of intensity – is antithetical to what is considered normal behavior for a female. Insofar as she experiences the feelings of the scientist, she is not feminine, and insofar as she accedes to the needs of her feminine nature, she won't be taken seriously as a scientist. Those childbearing years, when a woman is healthiest and has the most energy for child rearing, coincide with the peak opportunity years in any profession. And, to the extent that she is tempted to take a break, the cost to a woman's career, given the dominance of the male model in science, is high. Indeed, many of the reasons women give for leaving science and the reasons men give for not encouraging them to stay involve this double bind.

Where Millett's analysis particularly applies to women in science is in her consideration of the double bind. There are only three appropriate roles for women in our society: the mother role, the wife-like role (in science, the research associate), and the decorative role (only possible, incidentally, when you are young). If you don't fit naturally into any of those, you've bought a fourth option which belies description.

Careers considered appropriate for women are extensions of these three roles. Working with children or old people in some nurturing capacity is, of course, an extension of the mother role. Women who are research associates all their lives, or secretaries, or assistants are playing out a wife-like role for those they serve. And women who entertain, in every variety of that function, are decorative objects.

11.1.4 Youth and Genius

Within the ideology of science one additional bias is at work that we don't see in other professions: This is the powerful idea that any really good work must be done when the scientist is young. If a woman returns to a university to do science at 35, even if she intends to spend the next 30 years at it full time, her colleagues tend to believe she's unlikely to make a major contribution. Science as a young person's game is a myth that originates in data culled from 18th- and 19th-century science when men (and women) did not live very long. Most likely, according to a number of historians of science, another variable is at work besides the number of brain cells in youth: newness to the field. (Kuhn 1970; Gillmor 1984).

In the end, of course, knowledge is power, and so it is not surprising that any dominant group will try very hard to keep subordinate groups away from knowledge. We know that during slavery and beyond, African-Americans were forbidden to learn to read, the first tool of knowledge. In the same spirit, every colonial power has kept its "natives" to a limited educational avenue. As women seek the very highest achievements in their knowledge gains, it will bring them not only status but power, and that is exactly what the entire patriarchal structure is designed to prevent.

11.1.5 The Third Gender

There have always been women scientists. How did they deal with such traditions and proscriptions? Mostly, as Margaret W. Rossiter, Evelyn Fox Keller, and Vivian Gornick have amply documented, by accepting the restrictions and accommodating their ambitions to men's needs for domination (Rossiter,1982, Keller 1985, Gornick, 1990). Until the new wave of feminism about half of a century ago, the survival strategy of the typical American woman scientist was to persuade the men who taught her, funded her, and with whom she worked, that there were three sexes – men, women, and me. "What must be true of women in general is not true of me," such women asserted. To prove that to you, I will make myself as much like you, the dominant sex, as I can. I will deny my sisterhood with other women if that is the price I have to pay, deprive myself of family, if that is necessary. I'll have no spouse, pretend to have no social life, and certainly not display my sexuality. You may safely conclude I am not like other women, and therefore I don't deserve a female's status.

There is documentation for this pattern of accommodation, which some call internalized sexism and others the Queen Bee Syndrome. Rossiter, the historian of American women in science, tells us that women scientists were thankful for the tradition of using initials, not first names, on research papers, so they could appear as much like men (that is to disappear as women) as they could. Rossiter reports that many such achievers were reluctant to answer questionnaires about being women in a man's world. Those who were married followed their husbands as research associates, happy to be able to do science at all.

When Gornick interviewed 100 women scientists in the 1970s, some of them by then in their 60 s and 70 s, most of them had, they reported to her, had "good lives." It did not occur to them that they might have had better lives, might have done science at a much higher level, had they been willing to fight for what they deserved. They had not organized; they had not made waves; they had not complained. Such women might do science, even good science, but – and this is my interpretation – being "in denial" of how their own lives and work had been affected by their gender, they could not have been good mentors for young women. They had internalized the values of their oppressors, feminists would say.

Like men, such female scientists would most likely concur that, on average, women were probably not as good scientists as men – they, of course, being the exception. Such women would probably have had as little time and tolerance for gender-specific role conflict as the men with whom they worked. So a younger woman coming to talk about her need to schedule a pregnancy might find in older women scientists as deaf an ear as she would find in men. Of course, there were exceptions, but denial, historians say, was the norm.

When Keller's much heralded biography of the late Barbara McClintock was published, arguing that McClintock's insights into corn genetics were "feminine" in their challenge to the dominant command-and-control paradigm in mainstream cell biology, and that her isolation as a scientist was gender related, McClintock's response was, "Hogwash." She insisted that there was nothing in her work that had anything to do with maleness or femaleness. Despite neglect and bad treatment from the scientific establishment, she continued to internalize the values of the dominant class to the end of her life.

Feminism has been a force in American politics for 35 years. Instead of denying her womanhood, today's woman is willing to associate with other women and she knows that the men-women-me strategy must fail. But, as she pursues her scientific career, she runs into unregenerate views of what science is and what makes a good scientist that are held by most of her male mentors and colleagues – views virtually unchanged from the 1960s.

Many well-intentioned graduate professors, lab directors, and deans can and do point with pride to their fairness in handling their women students and to the absence of sexism, chauvinism, and sexual harassment in their domains. Individually, they attend to such issues because, superficially, they are liberated. But when a woman scientist fails, or quits, or doesn't achieve her predicted potential, they still blame her entirely. They are still unwilling to examine their behavior or her failure in terms of the prevailing norms (in politics we call these "belief systems") in science.

11.1.6 In Pursuit of "Excellence"

Why improve perfection? Those seeking to open science, to feminize it, will face serious opposition. Tens of thousands of scientists don't think science could be any better than it is. When then Radcliffe College president Linda S. Wilson, herself a chemist, in a talk at the National Academy of Sciences criticized the competitiveness and "fierce rivalries" in today's science, scientist-respondents to the report of her talk in The Scientist (January 20, 1992) wrote, at times in an angry tone, that they were merely defending "excellence." Such entrenched scientists think the quality of American science, at least at the graduate level, is the best in the world – educationally, the best it can be. Their approach, which is unashamedly elitist, means that only a very few can do science; hence, the educational process must be selective, must weed out all but the very fit in first-year college courses, graded on a curve.

Predestinarianism – a term I borrow from 17th century Protestant theology – also stands in the way of reform. Many scientists believe that people who are going to do science are born, not made, and will be discovered early, if at all. That's the reason why, when scientists get interested in education, they tend to tinker with the schooling of the very young. Why this skew? Because, many scientists believe, only educational investment in children is likely to pay off.

The third component of the dominant ideology that resists change in the practice of science is the idea that science, unlike most other professions, is a calling. As a result, issues of mobility and family needs are necessarily secondary to the work itself. This tradition not only inhibits women; it also weighs heavily on many men. A recent Ph.D. in chemistry told his chairman he didn't want to leave town right away because his wife was moving ahead in a banking job. His professor's response was to cease to help the young man find a postdoctoral position, stating: "If you're not going to take your career seriously, I don't have to." This professor spoke entirely from ideology. He had no data that men who are married to independent wives lead more productive lives in science than men who are not. His unreflective view was that a scientist must focus – whatever the personal cost – continuously on his career.

Linked to all these barriers to change in the practice of science is solipsism, the tendency to find truth and inevitability in one's own experience. The men in science – the professors and the mentors – are going to extrapolate from their own experience in helping women design theirs. With the best of intentions, they'll say, "When I was 28, I..." or "After my postdoc, I..." all of which may or may not be relevant for a woman. In the best circumstance, the woman can educate them, and explain to them why this particular career line isn't appropriate for her. In the worst possible case, as with the young chemist just described, they may think less of the student who is not moving in familiar tracks.

None of these conservative forces – elitism, predestinarianism, science as a calling, or solipsism – is particularly directed against women. But all four philosophies have disproportionately negative effects on those whose lifestyle or values or expectancies do not mirror these specific ideologies.

11.1.7 The Purpose: To Change Scientific Dogma

It is as difficult to argue someone out of a set of beliefs that have worked very well for their colleagues and themselves as to persuade someone to change religions. The feminist strategy since the late 1990s, therefore, shifted from an argument about means to a focus on outcomes.

In the first stage of our drive for equity, feminists attempted to achieve equality of opportunity or fairness, which we assumed to be synonymous with impartiality. We thought that if we could just remove structural barriers – male privilege, segregation, and gender bias – we would achieve educational equity. Later, it became clear that, even without barriers, women and men have different experiences in school. Minority students all the more. Teachers are influenced by stereotyping, latent status, gender, skin color, and ethnic origin.

Feminist activists and researchers soon learned that removing barriers would not suffice. It would be necessary somehow to achieve equality of experience. How? Some approaches include all-girls' math classes, compensatory training, career awareness, consciousness raising, and an extra dollop of math/science self-esteem. The test of whether there had been equality of experience would be whether there were equality of outcomes.

Today, the feminist educator's position is this: It does not matter to us how an institution gets there, what rules it adopts, or what rules it changes. We want to see women equally represented among the math/science majors and in the math/science professions. If, as some longitudinal studies of gifted children show, eight percent of the top one percent of math-achieving boys attain the Ph.D. in math, science, or engineering (Lubinski and Benbow 1992), then we want to see eight percent of the top one percent of math-achieving girls do the same. Whatever the individual behavior, we're looking for aggregate outcomes. So, for every female who drops out, there has to be some female who is attracted in.

11.1.8 The Benefits for Humankind

Whatever we do to enhance the attainment of women in science is most probably going to benefit men as well. We've already seen evidence that those changes that make women feel more comfortable in math and science – personal attention in a more collaborative atmosphere – help men as well. In response, many feminists are shifting focus from individual differences to a more organizational perspective.

We are no longer asking women to adapt themselves to existing structures, but to negotiate from strength for changes in established institutions and places of work. This means feminist scientists ought not look for palliatives but for more radical action. To make a lasting impact, young women in science must embrace feminism and acknowledge both their debt to the feminists awakened in the 1970s and their obligation to serve similarly the needs of the next generation of young women. There is no question in my mind that this country would not have moved off the male-dominant model had we not forced open doors during the past two decades. Whether a woman in science sees herself as "political" or not, out of a sense of obligation, or a wish to change history, all have a stake in the struggle.

There's a terrible staple of tradition in American feminism as documented by Susan Faludi: every 75 years or so, we have to start over from scratch (Faludi 1992). Our movement disappears from the history books and from our consciousness. For our sake, for the sake of our younger sisters, we must make sure that this does not happen again.

11.2 Tacit Discrimination and Overt Harassment: The Toll on Women, Minorities and the Nation

Linda S. Wilson, Ph.D., trained as a chemist, is president emerita of Radcliffe College, having led that institution from 1989 to 1999. Over the course of her career, she has served on numerous boards and committees established by the federal government, education associations, and the National Academy of Sciences. She is currently a senior lecturer at the Harvard Graduate School of Education.

Mentoring students and colleagues is one of the most important responsibilities of those whose talents and opportunities have enabled them to achieve. Mentoring should be much more than just offering information on how the "system" works and sharing personal experiences, as important as those contributions are. Mentoring should also include trying to improve the institutional structure and environment for those whose careers are still developing. Such comprehensive mentoring is an essential investment in our nation's future.

The nature of the economic base of this country presents a growing need for skilled and knowledgeable workers, for intellectual talent, and for ingenuity. How well we identify, develop, and encourage talent and how well we align talents with important tasks are among the core determining factors for our national and global wellbeing. These activities require long-sustained effort and a deeply ingrained understanding that talent is broadly distributed in the population, not stratified by gender, race, or socioeconomic status. This broad distribution of talent is a critical national resource, but it is one that has been far from adequately recognized or drawn upon. This new century will require better stewardship of this national resource.

Good work toward such stewardship has begun. Substantial progress has been made in changing the rules that previously barred women, minorities, and economically disadvantaged persons from entering education, the workplace, and many social environments But rule changes alone cannot dissolve the obstacles that limit their participation. Much work remains to be done.

Our nation's future economic security will depend on how well we continue to dismantle the obstacles to full and effective participation in society by all people. Women and minorities represent an important source of renewal for the occupations from which they have historically been excluded. As new entrants they bring probing and fundamental questions, fresh ideas, new energies and skills, and innovative and different perspectives to old problems. They are not blinded by the familiar or trapped by the status quo.

Furthermore, many of the challenges we now face will not yield just to knowledge and innovation. We need to develop, now more than ever, an informed, scientifically literate citizenry in all walks of life. Progress in the development of viable policies for energy, education, health care, the environment, and national security requires both well-conceived policies and concerted public will, i.e. the broad and sustained commitment by the public to make short-term sacrifices where necessary for long-term gain. In the future we cannot succeed without the full participation of women and minorities, both in the development and in the support of effective policy and action.

These facts lead to some vital questions for science and engineering in the United States:

- Are the science and engineering enterprises open and hospitable to women, to men, and to all people, no matter what their race and ethnicity?
- Were we to evaluate the design and operation of these professions and courses of study, would we conclude that their intention is to make women and minorities full participants or to continue to treat them as guests or outsiders?

From the vantage point of one who has had many opportunities to observe the environment for women in so-called "nontraditional" fields and to review reports of studies of the experiences of women and minorities, let me share some of what I have learned.

11.2.1 Postsecondary Science for Women: What Welcomes and What Inhibits

There is mounting evidence that our postsecondary institutions are less than hospitable to women in important ways.

The University of Michigan Center for the Education of Women has a rich repository of information and studies (http://www.umich.edu/~cew/index.htm). The group reviewed studies on women in science and mathematics at various institutions and published the results of a study about the experiences of women and men in mathematics and science programs at a particular major research university. The latter focuses on what students believe enhanced or inhibited their decisions to pursue or not to pursue academic work and careers in math and physics. The section presented below summarizes, excerpts, or quotes findings in this report.

11.2.2 Results from the Literature Review

- Prestigious liberal arts colleges are especially effective in producing women who subsequently earn PhDs.
- Self-confidence affects whether men and women choose to go into science and math and may well be the distinguishing characteristic between the approaches of men and women to the latter.
- Even women who perform as well as men in science and math report lower confidence in their ability to do science and math, and women's confidence drops during the critical early years in college.
- Women students must have unusually strong confidence in their abilities to counter prevailing notions that science and math are inappropriate fields for women.
- Women are especially sensitive to cues from the environment that reflect the quality of their performance and the likelihood of their success.
- Studies of classroom dynamics suggest that men and women have quite different levels of positive experiences. Women receive less direct encouragement, are taken less seriously, are given fewer opportunities, and are less likely to enjoy a comfortable relationship with counselors and teachers.
- The negative classroom dynamics seem to be worse for women majoring in mathematics and science. The women studied were nearly unanimous in believing that women receive less positive treatment. They report being "put down," patronized, called on less frequently, and ignored.
- Teachers who are particularly successful in encouraging women to pursue science and mathematics include information about women scientists in the curriculum, avoid gender-stereotyped views of science and scientists, and are sensitive to sexist language.
- Women generally respond negatively to what is perceived as an overly competitive environment. Typically, they find cooperative atmospheres more helpful and competitive atmospheres more harmful than do male students.
- Of the programs colleges have designed to increase the number of women who complete degrees in math and the physical sciences, very few focus on restructuring the academic environment or changing the discouraging attitudes and behavior of faculty members.

11.2.3 Results from Three University of Michigan Research Projects

At the University of Michigan, three research projects provided snapshots at varying temporal points along students' pathways to advanced education in mathematics and physical sciences:

(1) at the initial entrance into an accelerated undergraduate curriculum; (2) at graduation with a bachelor's degree in math or physics; and (3) in doctoral work in math or physics. The studies showed that:

- Academic performance and ability are clearly not what limit women's achievement.
- The critical factors are students' perceptions of science and math, of the educational environment, and of their own abilities.
- Encouragement, whether by faculty, family, or peers, appears to be the critical link in the chain that helps build and maintain the self-confidence needed to persist on the path to a higher degree in math or science.

11.2.4 The Stacked Deck Against Women in Science

Several major research universities have undertaken serious self-assessments of the academic environment for women in the sciences at the graduate student level and at the faculty level.

Some of those findings follow:

- With few exceptions, women in science are but a small minority in their peer groups, and their proportion drops sharply as they advance through their careers. The resulting isolation impedes research, increases stress, and may lead to abandonment of the profession.
- The period when successful scientific careers are usually forged (i.e. in the general career pattern developed when the enterprise was almost exclusively male) corresponds to the time of childbearing.
- Experimental work, which makes extraordinary demands on availability in time and/or location, raises conflicts with the family responsibilities that continue to be disproportionately borne by women.
- Women graduate students are often dissuaded from pursuing certain areas of science. In some disciplines, they are discouraged by faculty and student colleagues from pursuing mathematical or theoretical investigations; in other fields, women are discouraged from pursuing experimental work, especially where it involves fieldwork in remote geographical areas.

More subtle forms of discrimination also continue, including, for example, treatment of women as outsiders and negative attitudes of faculty toward women's family commitment. The predominance of males among the tenured science faculty in major research universities results in many nontenured women feeling powerless. Many nontenured women do not have positive, collegial relationships with senior members of their departments; they are deprived of the mentoring relationships often critical to advancement in a field. In some cases, junior faculty feel exploited and, in many instances, women perceive their situations to be worse than those of their male junior colleagues. The failure to integrate junior faculty into their own academic departments has been a long-standing problem (Fox 1993).

Some of our most distinguished research universities have recently uncovered significant inequities in the salaries and research facilities and resources provided for women faculty in comparison to those provided for men faculty. Even as these institutions take steps to remove inequities and to be more open in their recruiting of faculty, the number of women on the senior faculty in the science and engineering fields in these institutions continues to be extremely low.

Twenty years ago, a study by Louise F. Fitzgerald, Lauren M. Weitzman, Yael Gold, and Mimi Ormerod (Fitzgerald et. al. 1988) revealed that the power dynamics involved when faculty engage in sexual relationships with students, whether graduate or undergraduate, were still widely misunderstood and that their negative impact was seriously underestimated. During the intervening years many institutions have established policies on sexual harassment and unprofessional conduct. Many have procedures for resolution of complaints. Overt sexism is less pervasive than in the past. Nevertheless, serious instances of sexual harassment and gender-specific academic harassment continue to occur and add to the burden of the other obstacles women entering "non-traditional" fields must overcome.

In the last two decades, many scholars and institutions have undertaken studies of access to and the environment for women in higher education generally, and in science and engineering in particular. These efforts are commendable, but the studies frequently show that the barriers and problems that were found in similar reviews 30 years ago still persist. We need to move vigorously beyond just identifying these obstructions to devising and implementing ways to remove them.

Similarly, in the last decade or so, many studies have addressed workplace issues broadly, beyond the science and engineering fields. Some leading organizations, both corporate and academic, have introduced more "family-friendly" policies that recognize the shared obligations of parenthood and introduce various forms of flexibility and access to informational and financial resources for addressing them. These new policies need to be much more widely implemented. They also need to be supported by attitudinal change both in the workplace and more broadly in society if these improvements are to provide the relief that is needed and unfetter the talent on which we depend.

11.2.5 Addressing the Odds for a New Workforce

Most of these problems described above reflect a culture that has not sufficiently recognized the capabilities and contributions of women or valued their potential. They reflect a culture that has kept pace neither with women's changing skills and employment patterns nor with society's increasing need for women scientists' and engineers' talents. These problems are one result of our tendency to imagine the ideal scientist as a man who can single-mindedly and consistently devote 60–80 hours a week to science because he either has no conflicting familial obligations or places such commitments at a low priority. They also reflect a time when the nation needed far fewer scientists and engineers than is the case today.

Such a picture is now anachronistic. The assumptions on which it is based are not valid. Women constitute a substantial proportion of the workforce. Because two-parent families are now most often supported by multiple wage earners, the family support system at home, which previously permitted men in the workforce to concentrate their energies on work with intensity and without distraction, is now much less available to men and rarely available to women. For single women, whether or not they are parents, the support system for life responsibilities outside of the workplace is negligible, yet these individuals continue to have substantial cross-generational care giving responsibilities. Single parents, whether women or men, and whether widowed, divorced, or never wed, have great difficulty reconciling the expectations of their workplaces with the reality of their child care responsibilities. As the population in general ages, the elder care responsibilities of men and women in the workforce are also escalating with far from adequate institutional and community response.

In science and engineering, and in other professions that involve lengthy education and "apprentice" periods before the participants function fully and independently, these complexities must be endured for a long time and often with low compensation. Only the most highly motivated persons will brave this path as it has been designed and supported.

The higher demand for scientists and engineers today, the changed composition of the workforce, and the new reality of the situations workers face yield a substantially altered circumstance. Institutional designs, workplace policies, and attitudes of those in leadership and supervisory positions that fail to accommodate this altered circumstance cannot yield the quality and productivity the future requires.

Curiously, the United States is the only industrialized nation in which basic workplace policies still largely assume that women are not present. At the same time America educates its women far more than do other nations. Women's talent, education, and productivity represent major advantages for this country, advantages we are squandering by clinging to old assumptions about women's ability, about appropriate gender roles, and about how to teach and develop talent.

The specific studies referenced above focus primarily on the development of women's talent in science and engineering and are just a few among many. In addition, there are many important studies of the obstacles faced by minorities in these fields and more broadly. Many of the obstacles minorities experience are similar to those women face; others are even more severe.

All these studies lead to the conclusion that our assumptions reflect too inadequate an understanding of our national talent pool, too limited a view of the path to achievement, and too narrow a set of assumptions and criteria for identifying and assuring the excellence we desire.

11.2.6 A Mandate for Change

Many of the barriers to achievement have been identified. The dilemmas, the conflicts, the competing claims on human, physical, and financial resources are real and difficult. But our working and learning environments can be changed, and they need to be changed both for the sake of the women and men who study and work in them and for our society's future well-being.

Re-examination of the assumptions on which institutional designs are based, in light of current economic and social realities, should therefore be a high priority. The new environments we design to develop and nurture the talent we need can eliminate the discouragement and harassment women and minorities now experience. Their resulting self-confidence will also reduce their vulnerability to harassment. Improvements in the environment for women and minorities in science and engineering will benefit men as well and will broaden the base of talent that can contribute successfully in a knowledge-intensive society.

Change is a process, rarely just an event. This means that even with our best efforts, the transformation will evolve; even as inequities fade and obstacles diminish, they will still have to be faced. It is hard to be in a changing environment as a member of a disadvantaged group. It is hard to be positive about improvements while some serious difficulties persist. However, that is an inevitable part of social change. Those most affected will need to welcome and acknowledge the improvements even as they keep up the pressure for change where obstacles and inequities persist. Opportunities offered need to be exercised to ensure their persistence. Those who risk themselves helping others advance need to be appreciated and encouraged. Progress needs to be celebrated even as the tasks remaining lie ahead.

What is needed now are:

- A broader vision of what makes an enterprise productive with a special focus on science and engineering enterprises;
- Leaders to signal the value of changes in our assumptions and attitudes about the talents of women and minorities in general and in these fields;
- Innovative strategies to preserve core values of fairness and excellence as we transform the workforce and workplace for this new century;
- Widespread sharing of innovations that show promise;
- Mentors and their students to seize the opportunities as they emerge from new and more effective efforts to recruit and respond to their talents.

Out of these ingredients we should be able to forge a winning collaboration for constructive change.

11.3 The Red Shoe Dilemma

Lynn Margulis, Ph.D., Distinguished University Professor at the University of Massachusetts at Amherst in the department of Geosciences, has written over 300 scientific articles and 12 books. She has developed numerous hands-on teaching activities for levels from middle-school to graduate school.

For as long as I can remember, when someone asked me what I wanted to be when I grew up, I always answered "an explorer and a writer." Explorer of what? As a child, I didn't know: undersea cities, African jungle pyramids, unmapped tropical islands, and polar caves. "Whatever will need exploring," I said without hesitation. Today, nearly incessantly, I explore with passion the inner workings of living cells to reveal their evolutionary history. And, as soon as I learn something new about motility proteins, bacteria or insect symbionts that helps explain the early history of life on the Earth's surface, I write about it.

So you see, I am, after all these years, an explorer and a writer. Science for me is exploration, and no scientific work is complete if it has not been described and recorded in an article by the scientist herself (the "primary literature") or in a book or paper by someone else (the "secondary literature"). Much of my day is spent in description: Generating literature that speaks to fellow scientists and graduate students, talking in classes or lecturing to amuse the curious, writing notes and observations, collecting references, and jotting down the insights of others. I have become a mother (four children), a wife (twice), and a grandmother (five times, so far).

Because no one in my early life ever even explained the existence of scientific inquiry, I never realized until adulthood that I could participate in the great adventure of science as a profession. Unlike many friends, neither as an adolescent nor as a young adult did I wait for "my prince to come." Rather I expected some – any – opportunity to join serious expeditions. Then, as today, I read nearly everything in sight: scientific papers, bottle labels, train schedules, recipes, Spanish poetry, and novels. Decades ago, on the south side of Chicago, I used to ride the "IC" (Illinois Central Railroad) some 40 minutes, both in the stifling heat of summer and the freezing cold of winter, at least once weekly to the downtown "Loop" for ballet. Ballet classes (demanding, exhausting, French, and irrelevant) were sufficiently escapist to be captivating before scientists or exploratory missions were available in my life.

11.3.1 Choices

One film moved all of us dancers in those days; we all idolized red-headed Moira Shearer prancing in her Red Shoes. Set near Nice on the Mediterranean, close to a place with a marine station (Villefranche-sur-Mer) that I would get to know many years later, this romantic movie mesmerized my dancing classmates. The talent of this beautiful ballerina in the prima donna role was exhilarating, as was her true love for her sexy, handsome musician beau. I remember anger at the melodrama of that film, however. I thought the dichotomy of her life that led to her self-instigated fate utterly ridiculous.

Why did there have to be the "necessity to choose" between devotion to a man or a career? What generated the psychic dissonance that drove her to destruction? Obviously there was no reciprocity: If the star had been male, he would not have been driven to choose. He simply would have taken a wife. Instead, under relentless pressure to be the perfect dancer whose shoes run away with her, the ballerina yields to the dance master's demands that she remain in the spotlight, stage center of his world. But, equally enamored of her man, she is driven by another exigency: Her lover demands that she marry him and have a family.

Caught between the pressure of career and love, she could only resolve the conflict by suicide. Why hadn't she simply married her lover, borne their children, and continued dancing? Hollywood resolved her dilemma tragically, making the young heroine jump to her death from the summit of a sea wall. What infuriated me was the idea that the healthy, beautiful, and ambitious ballerina had to accept the "either-or" notion imposed upon her by the two men who ran her life. Should she simply have opted for everything, however, she would have deprived the film of its trumped-up fatal conflict. Wasn't a strong family life and a career possible for Moira Shearer's character? Isn't such a full life even easier today in the age of food storage by deep freeze, the electronic mail system, the private automobile, the dishwasher, and the laundry machine?

11.3.2 Having It All? Hardly!

At age 15, I was certain that the ballerina died because of a silly antiquated convention that insisted that it is impossible for any woman to maintain both family and career. I am equally sure now that the people of their generation who insisted on either marriage or career were correct, just as those of our generation who perpetuate the myth of the superwoman who simultaneously can do it all – husband, children, and professional career – are wrong.

Today many students, especially women, ask me for enlightenment, how to combine successfully career and family. When they learn I have four excellent, healthy, grown children and never abandoned science even for a single day in over 45 years, they request my secret. Touting me as an example of an American superwoman, they label me a "role model" (a term I despise). But there is no secret. Neither I nor anyone else can be superwoman.

Aspiring to the superwoman role leads to thwarted expectations, the helpless–hopeless syndrome, failed dreams, and frustrated ambitions. A lie about what one woman can accomplish leads to her, and her mate's, bitter disappointment and to lack of self-esteem. Such delusions and self-deceptions, blown up and hardened, have reached national proportions. Rampant misrepresentation of feasibility abounds as everyone falls short of the national myth peopled with a happy family, educated children, and professionally fulfilled parents. Something has to give: the quality of the professional life, of the marriage, of the child rearing – or perhaps all – must suffer.

11.3.3 A Dangerous Myth

The unreality of such expectations, coupled with the gross inadequacy of our educational system – such as it is – often leads to despair temporarily relieved by mind-numbing drugs – couch potato television addicts, marijuana, whiskey, cocaine, or other escapes.

Each husband, wife, and child in this sea of false hope suffers the crushing pain of inadequacy. In the United States, we value the beauty and strength of youth but, as a culture, we disdain love for children as "touchy-feely" and denigrate home-making as trivial and unworthy. We marginalize or expel the elderly and ridicule life on communes. By no means are the homeless on the street the only ones without homes. Unwilling to care for out greatest resource and those in direst need – our infants and children – we, speaking through money, debase their instructors, despising the seriousness needed to acquire a fine education. Our culture ridicules the intellectual while it lauds corporate greed and acquisitive ambition. It tends to ignore or misunderstand science while worshipping technology over-production and marketing.

11.3.4 One Woman's Path

I have not in any way overcome these stresses or resolved these common problems. I have just ignored them, as if they were laws that do not apply to me. Looking beyond such social heartaches, I chose intellectual exploration as my way of life and allied myself with nonhuman planet mates, with the scientific quest, rather than devoting myself to an arbitrary integrity of family and patriotism.

And, of course, I never jumped off the ballerina's cliff; the thought of abandoning life itself has always been unthinkable. Be warned, though, I do not offer a recipe for personal fulfillment – the superwoman does not exist, even in principle.

Mine is the story of scientific enthusiasm and enlightenment coming to a foolish and energetic girl who turned down dates on Saturday night and who never watched television. The point is that I was willing to work. This is not a statement of advocacy, as no single answer or easy path suits every woman. Probably, I have contributed to science because twice I quit my job as a wife. I abandoned husbands but stayed with children. I've been poor, but I've never been sorry.

For a single woman to raise emotionally healthy children, enjoy a happy husband, and make original contributions to world class science I believe is not simultaneously possible. Yet women who feel the urge must be encouraged to pursue scientific careers. Such women need our help. If life does not pose its problems as melodramatically as a Hollywood movie, neither does it resolve them so cleanly or definitively.

Yes, women can, of course, be superb scientists, but only at great sacrifice to their social life and its obligations. Most critically, productive women and girls must be surrounded by supportive and loving men and boys. We all need a cultural infrastructure that respects the deep needs of our young children and older family members. Let us hope that the provision of such enablers as scholarship monies, family leave opportunities, enlightened health insurance programs, imaginative and indulgent day care for preschoolers, and afterschool play-programs will increase the probability that talented and determined women will contribute much more to the scientific adventure in the future than they ever have been able to in the past.

The way I've put into practice these ideas to bring the inquisitive values of science to curious people is by development of hands-on curricular materials accompanied by video footage of live organisms. We have developed three units over the past 40 years. Our programs are designed for teachers: we respect their judgment. Sensitivity to their own student's needs will determine the levels at which the units are taught. Our goal is to provide teachers, scientists and science administrators with deep scientific knowledge and a classroom choreography that works in flexible ways for them. These three units have been used with middle school children. But I've taught the units to college teachers, graduate school students, and even school administrators. Everyone, I think, is enlightened by understanding the connection between shelled fossil protists called forams and the search for oil, coal, and gas or incinerated trash, fungal growth and carbon dioxide of the air. The way in which large-number estimation is connected to how evolution works by symbiogenesis and natural selection is an idea that educated people need to deeply understand.

Even after successful development of fabulous teaching materials, some of which hark back to enormous government investment in education post-Sputnik (c. 1963–1970) by the National Science Foundation, we have great difficulties. I have never had direct financial support for this activity since I'm not affiliated with a school of education or a science equipment supply company. Production and distribution of our teaching materials is limited by the lack of a ministry of education in this country, our own reluctance to restrict the units to a certain grade or subject matter (such as Advanced Placement Biology or Seventh Grade General Science when they are appropriate for both). Our solution to the production and distribution problem has been limited. With Dorion Sagan, my eldest son, we founded Sciencewriters (www.Sciencewriters.org). This partnership, with no employees and no capital, works without funding on the goal of bringing science to a wider constituency: students, teachers, artists, designers, museum-goers, environmental activists and others. Besides these three teaching units we continue to publish accessible books in this vein such as "Garden of Microbial Delights", the "Illustrated Five Kingdoms: A Guide to the Diversity of Life" and "The Microcosmos Coloring Book".

The final suggestion I have for you, if you are a young woman who wants to participate in the great scientific enterprise, is beyond the obvious requirement to be maximally educated in the content, methods, and wisdom of your science itself. I suggest you examine your own activities in a historical and social context. Ask first who wants the product of your labors (a pharmaceutical company, an educator who develops science curricula, an environmental impact assessment regulatory agency, a weapons lobbyist). Then realize, please, that if you are not in a position to communicate your ideas and promulgate your scientific results or educational strategies you can expect frustration and stress that you should never take personally. That is, each one of us is embedded in a cultural matrix, a language that simultaneously retrains and enables our expression, a determinant climactic geographical ambience and a single point in time in both natural and human history. These environmental factors make us who we are and, in the end, have far more relevance to our accomplishments than any individual differences in gender, character, emotional make-up, or level of personal dedication to the task at hand.

Chapter 12
Resources

"If you want to be successful, put your effort into controlling the sail, not the wind."

Anonymous

12.1 National Organizations for and of Women in Science, Technology, Engineering, and Mathematics

Association for Women in Mathematics, Fairfax, VA 22030
www.awm-math.org
Association for Women in Science, Washington, DC 20005
www.awis.org
Graduate Women in Science/Sigma Delta Epsilon, Avon, MA 02322
www.gwis.org
National Center for Women and Information Technology (NCWIT), Boulder, CO 80309
www.ncwit.org
National Institute for Women in Trades, Technology & Science, Alameda, CA 94501
www.iwitts.com/index.html
Society of Women Engineers, Chicago, IL 60611
http://societyofwomenengineers.swe.org/
The RAISE Project, Washington, D.C. 20036
www.raiseproject.org/
Women in Engineering Program Advocates Network, Denver, CO 80210
www.wepan.org

12.2 Mentoring Resources

Association for Women in Science, Washington, D.C. 20005
www.awis.org
MentorNet, San Jose, CA 95128

www.mentornet.net
Mentoring and Coaching for Executive Women, Thunder Bay, ON P7C 5N5
www.2020executivecoaching.com/index.php
Mentor/National Mentoring Partnership, Alexandria, VA 22314
www.mentoring.org
Mentoring Networks
http://www.womenswork.org/girls/refs/mentor.html
National Academies Committee on Women in Science and Engineering, Washington, DC 20001
http://www7.nationalacademies.org/cwsem/
Network of Executive Women, Chicago, IL 60601
www.newonline.org/mentoring.cfm
Telementoring Young Women in Science, Engineering and Computers,
http://www.edc.org/CCT/telementoring (archived website)
Women in Science and Healthcare Network (WISHnet), Bethesda, MD 20892
http://wish-net.od.nih.gov/

12.3 Organizations with Special Focus on Equity in STEM

American Association for the Advancement of Science, Education and Human Resources Programs, Washington, DC 20005
http://www.aaas.org/programs/education/CareersAll/index.shtml
American Indian Science and Engineering Society, Albuquerque, NM 87119
www.aises.org
Commission on Professionals in Science and Technology, Washington, D.C. 20005
http://www.cpst.org/
National Action Council for Minorities in Engineering, White Plains, NY 10601
www.nacme.org
National Consortium for Graduate Degrees for Minorities in Engineering and Science, Washington, D.C. 20006
www.gemfellowship.org
National Science Foundation, Directorate for Education and Human Resources, Arlington, VA 22230
www.nsf.gov
Quality Education for Minorities Network, Washington, DC 20036
www.qem.org
Society for the Advancement of Chicanos and Native Americans in Science, Santa Cruz, CA 95061
www.sacnas.org

Society of Mexican American Engineers and Scientists, Houston, Texas 77092
www.maes-natl.org

12.4 Organizations with Focus on Equity for Women

American Association of University Women, Washington, DC 20036
www.aauw.org
Catalyst, Inc., New York, NY 10005
http://www.catalyst.org/
Center For Women Policy Studies, Washington, DC 20036
http://www.centerwomenpolicy.org/
Equality Now, New York, NY 10023
www.equalitynow.org
Era Campaign Network
www.eracampaign.net
Gender Action, Washington, DC 20009
www.genderaction.org
Girls Incorporated, New York, NY 10005
www.girlsinc.org
Institute for Women's Policy Research, Washington, DC 20036
www.iwpr.org
International Center for Research on Women, Washington, DC 20036
www.icrw.org
National Coalition for Women and Girls in Education, Washington, DC 20036
www.ncwge.org
National Committee on Pay Equity, Washington, DC 20001
www.pay-equity.org/
National Council for Research on Women, New York, NY 10005
www.ncrw.org
National Organization for Women, Washington, DC 20005
www.now.org
The Wage Project
www.wageproject.com
Wider Opportunities for Women, Washington, DC 20036
www.wowonline.org

12.5 Field Specific Resources

12.5.1 Aerospace

Women in Aerospace, Washington, DC 20002
www.womeninaerospace.org/

12.5.2 Agronomy

American Society of Agronomy, Women in Agronomy, Crops, Soils, and Environmental Sciences Committee, Madison, WI 53711
www.agronomy.org

12.5.3 Anthropology

American Anthropological Association, Committee on the Status of Women in Anthropology Arlington, VA 22201
www.aaanet.org

12.5.4 Astronomy

American Astronomical Society, Committee on the Status of Women in Astronomy, Washington, DC 20009
www.aas.org

12.5.5 Biology

American Institute of Biological Sciences, Diversity Affairs Committee, Washington, DC 20005
www.aibs.org
American Society for Cell Biology, Women in Cell Biology Committee, Bethesda, MD 20814
www.ascb.org
American Society for Plant Biologists, Women in Plant Biology Committee, Rockville, MD 20855
www.aspb.org
Ecological Society of America, Washington, DC 20006
www.esa.org
National Association of Biology Teachers, Reston, VA 20191
www.nabt.org
American Phytopathological Society, Joint Committee of Women in Plant Pathology and Cultural Diversity, St. Paul, MN 55121
www.apsnet.org/

12.5.6 *Biomedical Sciences*

American Association of Immunologists, Committee on the Status of Women, Bethesda, MD 20814
www.aai.org
American Institute for Medical and Biological Engineering, Women in Medical and Biological Engineering Committee, Washington, D.C. 20006
www.aimbe.org
American Physiological Society, Committee on Women in Physiology, Bethesda, MD 22201
www.the-aps.org
American Society for Biochemistry and Molecular Biology, Education and Professional Development Committee, Bethesda, MD 20814
www.asbmb.org
American Society for Bone and Mineral Research, Women in Bone and Mineral Research Committee, Washington, D.C. 20036
http://asbmr.org/
American Society for Microbiology, Committee on the Status of Women in Microbiology Washington, DC 20036
www.asm.org
American Society for Pharmacology and Experimental Therapeutics, Committee on Women in Pharmacology, Bethesda, MD 20814
http://www.aspet.org
Biophysical Society, Committee on Professional Opportunities for Women, Bethesda, MD 20814
www.biophysics.org
National Institutes of Health, Office of Research on Women's Health, Bethesda, MD 20892
http://orwh.od.nih.gov/career.html
National Institutes of Health, Office of Science Education, Careers in Science, Women in Research Poster Series
http://science.education.nih.gov/home2.nsf/Careers
Society for Neuroscience, Committee on Women in Neuroscience, Washington, DC 20036
www.sfn.org
Women in Biomedical Careers
http://womeninscience.nih.gov/

12.5.7 *Chemistry*

American Chemical Society, Women Chemists Committee, Washington, DC 20036
www.acs.org

COACh: Committee on the Advancement of Women in Chemistry, Eugene, OR 97403
http://coach.uoregon.edu/index.html
Iota Sigma Pi, National Honor Society for Women in Chemistry
www.iotasigmapi.info/default.htm

12.5.8 Computer Sciences and Information Technology

Anita Borg Institute for Women and Technology, Palo Alto, CA 94304
www.anitaborg.org
Association for Women in Computing, San Francisco, CA 94104
www.awc-hq.org
Computing Research Association, Committee on the Status of Women in Computing Research, Washington, D.C. 20036
www.cra.org
Women in Technology International, Sherman Oaks, CA 91423
www.witi.com

12.5.9 Education

Aspira Association, Inc., Washington, DC 20005
www.aspira.org
Association of American Colleges and Universities, Program on the Status and Education of Women, Washington, DC 20009
www.aacu.org
National Science Teachers Association, Arlington, VA 22201
www.nsta.org
Women's Research and Education Institute, Washington, DC 20006
www.wrei.org

12.5.10 Engineering

American Institute of Chemical Engineers, Women's Initiatives Committee, New York, NY 10016
www.aiche.org
American Society for Engineering Education, Washington, DC 20036
www.asee.org
Institute of Electrical and Electronics Engineers, IEEE Women in Engineering Committee, Washington, DC 20036
www.ieee.org

National Society of Black Engineers, Alexandria, Virginia 22314
www.nsbe.org
National Society of Professional Engineers, Alexandria, VA 22314
www.nspe.org
Society of Hispanic Professional Engineers, Los Angeles, CA 90022
http://oneshpe.shpe.org
Women's International Network of Utility Professionals, Fergus Falls, MN 56538
www.winup.org

12.5.11 Geography

Association of American Geographers, Committee on the Status of Women in Geography, Geographic Perspectives on Women Specialty Group, Washington, DC 20009
www.aag.org
Society of Women Geographers, Washington, DC 20003
www.iswg.org

12.5.12 Geosciences

American Geological Institute, Alexandria, VA 22302
www.agiweb.org
Association for Women Geoscientists, Lincoln, NE 68503
www.awg.org
Geological Society of America, Committee on Minorities and Women in the Geosciences, Boulder, CO 80301
www.geosociety.org
Society of Exploration Geophysicists, Geo-Mentoring Committee, Tulsa, OK 68503
www.seg.org
Women in Mining, Lakewood, CO 80226
www.womeninmining.org
Women in Natural Resources, Moscow, ID 83844
www.cnr.uidaho.edu/winr

12.5.13 Mathematics, Statistics, and Economics

American Economics Association, Committee on the Status of Women in the Economics Profession, Tallahassee, FL 32303
www.cswep.org

American Mathematical Society, Providence, RI 02904
www.ams.org
American Statistical Association, Committee on Women in Statistics,
 Alexandria, VA 22314
www.amstat.org
Black Women in Mathematics, Buffalo, NY 14260
http://www.math.buffalo.edu/mad/wmad0.html
Caucus for Women in Statistics
http://caucusforwomeninstatistics.com/
Institute of Mathematical Statistics, Beachwood, OH 44122
www.imstat.org
Joint Committee on Women in the Mathematical Sciences
http://www.csupomona.edu/~hlord/jcw/
Mathematical Association of America, Committee on the Participation of
 Women in Mathematics, Washington, DC 20036
www.maa.org
Math/Science Network, Oakland, CA 94613
http://www.expandingyourhorizons.org/
National Council of Teachers of Mathematics, Reston, VA 20191
www.nctm.org
Society for Industrial and Applied Mathematics, Philadelphia, PA 19104
www.siam.org

12.5.14 Medicine and Health

American Association of Women Dentists, Chicago, IL 60606
www.aawd.org
American Medical Women's Association, Philadelphia, PA 19103
www.amwa-doc.org
American Medical Association, Women Physicians' Congress, Chicago, IL
 60629
www.ama-assn.org/ama/pub/category/18.html
Hispanic Dental Association, Springfield, IL 62703
www.hdassoc.org
American Pharmacists Association ,Committee on Women's Affairs, Washington,
 DC 20005
www.aphanet.org
American Public Health Association, Committee on Women's Rights, Washington,
 D.C. 20001
www.apha.org
Association of Women Psychiatrists, Dallas, TX 75228
www.womenpsych.org/

Association of Women Surgeons, Downers Grove, IL 60515
www.womensurgeons.org
Kappa Epsilon Professional Pharmacy Fraternity, Overland Park, KS 66202
http://kappaepsilon.org/
Society for Women's Health Research, Washington, DC 20036
www.womenshealthresearch.org

12.5.15 Meteorology

American Meteorological Society, Board on Women and Minorities, Boston, MA 02108
www.ametsoc.org

12.5.16 Physics

American Association of Physics Teachers, College Park, MD 20740
www.aapt.org
American Physical Society, Committee on Status of Women in Physics, College Park, MD 20740
www.aps.org
American Nuclear Society, Professional Women in the American Nuclear Society, La Grange Park, IL 60526
www.ans.org
SPIE, Women in Optics, Bellingham, WA 98227
http://spie.org/x10.xml

12.5.17 Psychology

American Psychological Association, Committee on Women in Psychology, Washington, DC 20002
www.apa.org
American Psychological Association of Graduate Students
www.apa.org/apags/
Association for Women in Psychology, Salt Lake City, UT 84105
www.awpsych.org/
Association for Psychological Science, Washington, DC 20005USA
www.psychologicalscience.org/
Psi Chi, National Honor Society in Psychology, Chattanooga, TN 37403
www.psichi.org

12.5.18 Sociology

American Sociological Association, Committee on the Status of Women in Sociology, Washington, D.C. 20005
www.asanet.org

12.5.19 Toxicology

Society for Toxicology, Women in Toxicology Special Interest Group, Reston, VA 20910
www.toxicology.org

12.5.20 Veterinary Medicine

Association of American Veterinary Colleges, Gender Issues Committee, Washington, D.C. 20005
www.aavmc.org/
Association for Women Veterinarians Foundation, Denver, PA 17517
www.vet.ksu.edu/AWV/about.htm

References

Broad WJ (1981) The publishing game: Getting more for less. Science 211: 1139
Chin J (2008) The light at the end of my tunnel isn't a research job: a guide to Ph.D. career transitions. AWIS Magazine (Winter) 38(1): 38–40
Dovidio JF, Gaertner SL (1998) On the nature of contemporary prejudice. In: Eberhardt JL, Fiske ST (eds) Confronting Racism, the Problem and the Response. Sage, Thousand Oaks, CA
Faludi S (1992) Backlash: The undeclared war against American women. Crown, New York
Fitzgerald LF, Weitzman LM., Gold Y, Ormerod M (1988) Academic harassment: sex and denial in scholarly garb. Psychology of Women Quarterly, 12(3): 329–340
Fox MF (1993) Women, men, and the social organization of science. AWIS Magazine, 22(1): 17
Fox MF (1999) Gender, hierarchy, and science. In: Chafetz, JS (ed) Handbook of the Sociology of Gender. Kluwer/Plenum New York
Gillmor CS (1984) Aging of geophysicists. Eos, 65(20): 353–354.
Gornick V (1990) Women in science. Simon and Schuster, New York
Horvath MM (2006) Mentoring the postdoctoral condition. AWIS Magazine (Summer) 35(3): 15–17
Horvath MM (2007a) Coaching vs. mentoring: the difference is in the details. AWIS Magazine (Spring) 36(2): 20–22
Horvath MM (2007b) Effective mentoring to achieve a healthy work-life balance. AWIS Magazine (Summer) 36(3): 24–25
Horvath MM (2007c) Science-minded advice for the undergraduate health professional. AWIS Magazine (Fall) 36(4): 31
Keller EF (1985) Gender and science. Yale University Press, New Haven
Kuhn TS (1970). The structure of scientific revolutions (2nd ed., rev.). University of Chicago Press, Chicago
Lubinski D, Benbow CP (1992) Gender differences in abilities and preferences among the gifted: implications for the math/science pipeline. Current Directions in Psychological Science, 1(2): 61–66
Millett K (1970) Sexual politics. Doubleday, New York
Rossiter MW (1982) Women scientists of America (Vols. 1–2). Johns Hopkins University Press, Baltimore
Rosser, SV (2006) Senior women scientists: overlooked and understudied? J.of Women and Minorities in Science and Engineering, 12: 275–293.
Rosser, SV (2004) The science glass ceiling: Academic women scientists and their struggle to succeed. Routledge, New York
Ruskai MB (1991) Comment: Are there cognitive gender differences? Am. J Physics, 59(1): 11–14
Slapikoff S (1985) Teacher's unconscious sexism improved. Radical Teacher, 7(28): 33

Sonnert G, Holten G (1996) Career patterns of women and men in the sciences. American Scientist 84: 63–71

Steinem G (1992) Revolution from within: a book of self-esteem. Little Brown, Boston

Suiter MJ (2006) Wisdom on mentoring: sharing the methods of exemplary science and engineering mentors. AWIS Magazine (Winter) 35(1): 17–25

Widnall SE (1988) Voices from the pipeline, American Association for the Advancement of Science Presidential Lecture. Science 241: 1740–1745

Valian V. (1998) Why so slow? The advancement of women. MIT Press, Cambridge, MA 1998

Bibliography

Association for Women in Science (2005) A hand up: women mentoring women in science. The Association for Women in Science, Washington, D.C. http://www.awis.org

Center for the Education of Women, University of Michigan, Ann Arbor http://www.umich.edu/~cew/research/respubs.html

Committee on Science, Engineering, and Public Policy, National Academy of Sciences (1997) Adviser, teacher, role model, friend. National Academy Press, Washington, D.C. http://www.nap.edu/catalog.php?record_id = 5789

Horvath MM (2008) Mentoring resources for science professionals. AWIS Magazine (Spring) 37(2): 32

Johnson A (2006) Minority women in science. AWIS Magazine (Fall) 34(4): 9–11

Lee A, Dennis C, Campbell P (2007) Nature's guide for mentors. Nature 447: 791–797

Oransky I, Harding A (2005) Diversity in the life sciences. The Scientist, November 7 Supplement: 1–54

Perlmutter DD (2008) Are you a good protégé? Chronicle of Higher Education (Careers): C1–C2

Index

A
AAAS, *see* American Association for the Advancement of Science (AAAS)
Advisor, definition, 4
American anomaly, 74
American Association for the Advancement of Science (AAAS), 44
American feminism, 136
"Antimentors", 11
Association for Women in Science (AWIS), 23, 25, 26, 39, 47, 71, 73, 81, 120, 126
AWIS, *see* Association for Women in Science (AWIS)

C
California Alliance for Minority Participation in Science (CAMP), 75
Career and life transitions
  coaching or mentoring
    embracing new communication paradigms and developing priorities, 56–57
    help with procrastination, 56
  transitioning into different career pathway
    create opportunities, 59
    key questions for change in career pathway, 57t
    knowing strengths/thought processes/values, 58
    let go of what one wants, 58
    making scientific background work, 58
    price tag on procrastination, 59
  work–life balance
    managing employer's expectations and self, 52
    strategies to attain balance, 53–54
Career success, moving toward
  gender and diversity issues, 87–92
    female anti-feminism, 91
    isolation, 80
    queen bee syndrome, 88
  leaping barriers and achieving goals, 83–85
  making connections, 81–83
  staying the course, 93–94
  timing and choices, 85–87
    put off having children, 85
    spouses, support of, 85–86
Children, put off having, 85
  see also Career success, moving toward
Coaching or mentoring
  help with procrastination, 56
  new communication paradigms and priorities, development of, 56–57
  see also Career and life transitions
"Comfort zone", 22
Connection/confidence, building of, 81–83, 112–113
  building connections, recommendations, 117–118
  claiming intelligence/confidence and femaleness, 115–117
  dyad format, 116
  sexism/internalized sexism/stereotypes about scientists, 113–115
  towards gender equality, 113, 118
Co-op college programs, 122
Cyberspace, mentoring in
  mentoring in new era of social media, 41–43
  Ph.D. Career Clinic, 44
  science careers forum, 44
  using web 2.0 technology to empower specific groups, 43

D

Discrimination and harassment, 135–136
  addressing the odds for new workforce, 140–141
  choices, 143–144
  dangerous myth, 145
  mandate for change, 141–142
  one woman's path, 145–147
  postsecondary science for women, 137
  red shoe dilemma, 142–143
  results, three university of Michigan research projects, 138–139
  results from literature review, 138
  stacked deck against women in science, 139–140
  see also Thoughts, for future
Dual-career marriage, 86
Dynamics and needs, changing
  life-long mentoring
    at all post-training career levels, 48
    at postdoctoral level, 46–48
    at student and trainee level, 45–46
  mentoring for under-represented groups
    mentornet, 39–40
  mentoring in cyberspace
    in new era of social media, 41–43
    Ph.D. Career Clinic, 44
    science careers forum, 44
    using web 2.0 technology to empower specific groups, 43

E

Experience, voices of
  applying for fellowships/research grants
    graduate school/postdoctoral fellowships, 105–107
    research grants, 107–109
  building confidence and connection, 112–113
    claiming our intelligence, confidence and femaleness, 115–117
    dyad format, 116
    recommendations for, 117–118
    sexism/internalized sexism/stereotypes about scientists, 113–115
    working toward gender equality, 113, 118
  graduate school and beyond, keys to success in
    choosing an advisor, 110
    conversation with lab director, 110
    final tips, 112–113
    joining lab, 110–111
    meeting with thesis committee, 112
  helping those who follow
    girls, guidance for, 119
    guidance, quality of, 121–123
    opening doors for girls, 120
    thinking globally, 121
  professional responsibility
    helping women entering science, 125–126
    literacy and expertise, 124
    serving society, 126–127
    working with colleagues, 126
  things to be told by professor
    about extracurricular activities, 102
    avoid common pitfall, 103
    be a good mentor, 103
    finances, 103
    gaining opportunity/equality/power, 98–100
    learning academic structure, 100–101
    make work visible/known/valuable, 103
    once in power, 103–104
    starting off right, 101–102
    women faculty, empowerment, 104
  women speakers
    memorize introduction/conclusion, 96
    practice, importance of, 98
    talking to audience, 96–97
    time limit, 97
    use of visual aids, 97–98

F

"Family-friendly" policies, 140
  see also Thoughts, for future
Fear, 59
  see also Career and life transitions
Fellowships or research grants, applying for
  graduate school/postdoctoral fellowships, 105–107
  research grants, 107–109
  see also Experience, voices of
Female anti-feminism, 91
  see also Career success, moving toward
"Feminism 101," 88
"Finding Your First Industry Job," 44
Formal mentoring, advantages, 47–48

G

Garden of Microbial Delights, 146
Gender, third, 132–133

Gender and diversity issues, 87–92
　female anti-feminism, 91
　isolation, 80
　queen bee syndrome, 88
　see also Career success, moving toward
Graduate education, 76–78
　academic assistantships for science students, 76
　choices, importance of, 77
Graduate school
　choosing advisor, 110
　direct conversation with lab director, 110
　joining lab, 110–111
　meeting with thesis committee, 112
　and postdoctoral fellowships
　　allow plenty of time, 107
　　contact funding agency, 105
　　identify possible fellowship sources, 105
　　plan ahead to arrange reference letters, 105–106
　　provide context/detail in research plan, 106
　　relate research plan/plan of study to career goals, 106–107
　　request current program guidelines, 105
　　seek review/comment on application, 107
　　submit application, neat/format guidelines, 107
　see also Career success, moving toward; Experience, voices of

I
Illustrated Five Kingdoms: A Guide to the Diversity of Life, 146
Informal education, 72–73
Internalized sexism, 113–116, 132
International Coach Federation guidelines, 55
Interpersonal contexts
　defining issues, 63–65
　　for institutions, 65
　　for mentors, 65
　　for protégés, 64
　　for protégés and mentors, 64
　identifying problematic behaviors, 62–63
　　checklist of questions, 63
　inappropriate relationships with mentors/supervisors, 61–62
　intervention strategies, 66–69
　　see also Interventions, offensive behaviour

Interventions, offensive behaviour
　direct response, 67
　humor, 66–67
　letter to perpetrator, 68
　pretending not to understand, 67
　private reprimand, 67
　records of observation, 69
　report, behavior, 69
　sending letter about issues, 68
　surprise, 67
　see also Interpersonal contexts

L
Les Aventures de Telemaque (François Fénelon), 3
Life-long mentoring
　at all post-training career levels, 48
　　"reverse mentoring", 48
　　soft skills development, 48
　at postdoctoral level, 46–48
　　formal mentoring, advantages, 47–48
　　participation in professional society activities, 47
　at student and trainee level, 45–46
　see also Dynamics and needs, changing

M
Mentor, potential, 8
Mentoring, 5*t*
　definition, 3, 4*t*
　Greek mythology, 3
　long-term face-to-face, 122
　reverse, 48
Mentoring relationships
　make relationship work
　　being self and doing well in all interactions, 21–22
　　patterns in life/career, 22
　　recognizing that actions have consequences, 23 seek hidden/unwritten/inside rules, 23
　　sense of humor, 22–23
　mentoring impact
　　protégés wants/not want, 24
　　from protégés to mentor, voices from field
　　　age doesn't matter, 34–35
　　　e-mentoring to attain workplace success, 31
　　　finding positive place, 30–31
　　　many mentors, many perspectives, 28

Mentoring relationships (cont.)
   mentoring is colorblind, 32–33
   never too busy to help, 29–30
   power neutral mentoring, 33–34
   questions posed to mentors, 27–28
MentorNet, 31, 39–40, 52, 53
The Microcosmos Coloring Book, 146
"Miss Manners" approach, 64
   see also Interpersonal contexts
Myths about mentoring, 12t

N

Networking, 59
   see also Career and life transitions
"Networking 101," 44
Ninth International Conference of Women Scientists and Engineers (England), 119

P

People-professionals, definition, 130–131
Postdoctoral years, 78–79
   see also Right education, starting with
Postsecondary science for women
   discrimination and harassment, 137
Potential mentor, 8
Precollege education, 73–74
   see also Right education, starting with
Predestinarianism, 134
Preparing to be mentored
   identifying your mentoring needs
      acquiring requisite professional credentials, 7
      balancing pieces of one's life, 9
      calculated risk taking, 10–11
      creating opportunities for others, 10
      dealing with own biases/misconceptions, 8
      developing sense of one's career directions/timing, 8–9
      knowing when to move on, 10
      learning from mistakes/missteps, 8
      meshing one's values with workplace, 9
      recognizing opportunity, 7–8
      selecting appropriate role models, 9
   mentoring models
      myths about mentoring, 12t
      questions for protégés to contemplate at any time, 14, 14t
      terminating mentoring relationship, 13
   techniques/tools for starting mentoring relationship
      advice to be considered, 20t
      pick mentor and start the process, 17
      questions to be asked, 17–18
      roles in fostering opportunities/creating environments, 15
      "rules" to be followed, 18–19
      specific responsibilities as protégés, 16t
Problematic behaviors, identifying, 62–63
Professional advisory boards, 47
Professional responsibility
   helping women entering science, 125–126
   literacy and expertise, 124
   serving society, 126–127
   working with colleagues, 126
   see also Experience, voices of
Professor, things to be told by
   about extracurricular activities, 102
   be good mentor, 103
   finances, 103
   gaining opportunity/equality/power, 98–100
   learning academic structure, 100–101
   make your work visible/known/valuable, 103
   once in power, 103–104
   starting off right, 101–102
   women faculty, empowerment, 104
   see also Experience, voices of

Q

Queen bee syndrome, 88, 91, 132

R

Relationships, inappropriate, 61–62
   see also Interpersonal contexts
Research grants
   request current program guidelines
      allow plenty of time, 109
      be confident/persistent, 109
      contact funding agency, 108
      get appointed, ad hoc, 109
      not funded, revise/resubmit, 108–109
      prior to submission, seek comment on proposal, 108
      provide context/detail in research plan, 108
"Research instructorships", 78
Resources
   field specific resources, 151–158
      aerospace, 151

Index

agronomy, 152
anthropology, 152
astronomy, 152
biology, 152
biomedical sciences, 153
chemistry, 153–154
computer sciences and information technology, 154
education, 154
engineering, 154–155
geography, 155
geosciences, 155
mathematics, statistics, and economics, 155–156
medicine and health, 156–157
meteorology, 157
physics, 157
psychology, 157
sociology, 158
toxicology, 158
veterinary medicine, 158
mentoring resources, 149–150
National organizations for and of women in science, technology, engineering, and mathematics, 149
organizations with focus on equity for women, 151
organizations with special focus on equity in STEM, 150–151
Reverse mentoring, 48
Right education, starting with graduate education, 76–78
academic assistantships for science students, 76
choices, importance of, 77
informal education, 72–73
postdoctoral years, 78–79
precollege education, 73–74
undergraduate education, 74–76

S
Self-assessments of academic environment for women in sciences, 139
Sexual harassment policies, 64, 65, 68
Sexual issue in mentoring relationship, *see* Interpersonal contexts
Social media (Web 2.0), mentoring in new era of
audio blogging/podcasts, 42
blogging, 41–42
social networking, 42–43
tagging and social bookmarking, 42

see also Dynamics and needs, changing
Spouses, support of, 85–86
Supervisor, definition, 4

T
Television programs, 119
Third gender, 132–133
Thoughts, for future
discrimination and overt harassment, 135–136
addressing the odds for a new workforce, 140–141
choices, 143–144
dangerous myth, 145
deck against women in science, 139–140
mandate for change, 141–142
one woman's path, 145–147
postsecondary science for women, 137
red shoe dilemma, 142–143
results, three university of Michigan research projects, 138–139
results from literature review, 138
"problem" of women in science
benefits for humankind, 135–136
implications for women in science, 131
patriarchy, 129–130
in pursuit of "excellence", 133–134
scientific dogma, to change, 135
third gender, 132–133
upending traditions, 130–131
youth and genius, 132
Touchstones (three), 72
"Tough love", 103
Transition of career pathway
create opportunities, 59
key questions for change in career pathway, 57t
know strengths/thought processes/values, 58
let go of what one want, 58
making scientific background work, 58
price tag on procrastination, 59
see also Career and life transitions
"Two-for one", 38

U
Undergraduate education, 74–76
see also Right education, starting with

University of Michigan Center for the Education of Women, 137
"Urban myth," 37

W
Web 2.0, 41
Women in science, problem of
  benefits for humankind, 135–136
  implications for women in science, 131
  patriarchy, 129–130
  in pursuit of "excellence", 133–134
  scientific dogma, to change, 135
  third gender, 132–133
  upending traditions, 130–131
  youth and genius, 132
  see also Thoughts, for future
Women speakers
  memorize introduction/conclusion, 96
  practice, importance of, 98
  talk to audience, 96–97
  time limit, 97
  use visual aids, 97–98
Workaholism, myth of, 52
Work–life balance
  managing employer's expectations and self, 52
  strategies to attain balance, 53
  see also Career and life transitions